Praise for *A Practitioner's Guide to Defense Sector Reform*

Effective defense forces are at the core of every nation's sovereignty. In a world of sovereign nations, weak, failing, and failed states are a threat to local, regional, and international security. That is why the United States and other countries have invested so much in the development of foreign security forces—often in vain as the hoped-for reforms fail to take root. The problem, as Hanlon and Kerr note in this welcome practitioner's guide, is "what to do?" Hanlon and Kerr's book adds substantially to the growing literature on security sector assistance and defense sector reform, providing analytic insight informed by illustrative examples. Read it to familiarize yourself with the evolving concepts; study it to help calibrate expectations; and apply it to get the best possible results.

—*Michael Miklaucic,*
Senior Fellow, National Defense University

A Practitioner's Guide to Defense Sector Reform serves as an indispensable reference tool and handbook for policy practitioners seeking to generate sustainable change in the defense sector. The guide frames defense sector reform along ten goals—conceptualized as "assistant packages" that cover the broad spectrum of activities, principles, and concepts that contribute to real, measurable, and sustainable change in the defense sector. The diversity and range of these goals is testament to the guide's comprehensive scope, covering topics directly related to democratic governance, such as "democratic control" and "civilian control," to rather managerial areas such as "logistics" and "military effectiveness." Combining the why, what, and how of reform within each of these respective areas makes the book a worthwhile read. Furthermore, its authors do not shy away from difficult topics—they acknowledge the multitude of interwoven challenges and are acutely aware of the political dimensions of defense sector reform, a process rife with resistance, bargaining, and conflict. Indeed, for an improved understanding as to why some reforms succeed while others fail, it is essential to examine them through a political lens. Finally, the authors clearly differentiate between defense sector reform and military assistance—support delivered via training and equipment—which is unalike reform. In conclusion, this is a sobering guide for anyone who believes that defense reform is simple, linear, and a military-technical process void of politics. However, it is empowering and full of insights for those who want to take defense sector reform to the next level.

—*Hans Born, Assistant Director and Head of*
Policy and Research Department, DCAF—
Geneva Centre for Security Sector Governance

This guide makes a valuable contribution to practitioners who need specific guidance on how to conduct effective defence sector reform programming. My colleagues and I have had many recent conversations with people directly involved in the design and implementation of defence sector reform programmes and they have expressed frustration that they lack a guide on how to conduct this important and complex work. This guide meets this need and will serve as a catalyst to improve defence sector reform programmes.

—*Nicholas Marsh, Senior Researcher,*
Peace Research Institute Oslo (PRIO)

The transformation of any defense sector lies at the heart of improving the effectiveness, efficiency, accountability, and affordability of a country's armed forces. It is a process that should be undertaken within the context of a wider review of the security sector. Defense sector transformation (DST), or defense sector reform (DSR), is a nationally owned process and is usually driven by strategic shifts, political transformation, technological innovation, resource issues, and/ or social change. Successful leaders in the security sector recognize the strategic nature of DST/DSR, the importance of context and vision building within a strategic defense review, and the need for a long planning horizon. This new practical guide by Strategic Capacity Group provides invaluable insights and pointers for those that advise, assist, plan, and implement a DST/DSR program. It takes the reader right to the heart of defense transformation by providing a spectrum of strategic focus areas together with the "triggers and hooks" which policymakers and practitioners alike need in order to find the best start/ entry point and then to properly sequence further sustainable interventions.

— *Brigadier (Retired) Gordon Hughes, CBE*

A Practitioner's Guide to Defense Sector Reform

A Practitioner's Guide to Defense Sector Reform

Querine Hanlon and Alexandra Kerr

StrategicCapacity

Strategic Capacity Group
8401 Greensboro Dr, Suite 1000A
McLean, VA 22102
www.strategiccapacity.org

First published 2022

Printed in the United States of America

The paper used in this publication meets the minimum requirements of American National Standards for Information Science—Permanence of Paper for Printed Library Materials, ANSI Z39.48-1984.

Cover photo by Querine Hanlon.
Interior design by Nigel Quinney.

ISBN 979-8-9860250-0-1 (paperback)
ISBN 979-8-9860250-1-8 (e-book)

CONTENTS

FOREWORD

The U.S. withdrawal from Afghanistan in the summer of 2021 and the subsequent collapse of two decades of investment in a security architecture for the country raises the perennial question of how to "fix" security assistance and ensure that such investment produces a return. Clearly, the United States and its allies need more options for how security assistance is delivered. Fortunately, this timely guide responds to that need by addressing two of the biggest challenges practitioners face: determining *what to do* and deciding *where to start*.

Dr. Querine Hanlon founded Strategic Capacity Group and helped developed this *Practitioner's Guide to Defense Sector Reform* in order to bridge the conceptual worlds of the academic and policy communities with the field-based realities of practitioners. The vital contribution made by this guide is to help program managers and officers who are tasked with "doing something" to *better define what they can achieve and how to go about doing it*. To accomplish this goal, the guide draws not only on studies of East European countries joining NATO after the Cold War but also on lessons from today's foreign assistance frontlines. As I learned from my own experience in standing up an irregular warfare center that brings together academics and practitioners, when time frames are compressed, funding cycles are short, and priorities are urgent, practitioners need frameworks that are elegant, easy to communicate, and actionable. This *Practitioner's Guide* delivers all three with gusto.

At the heart of the guide is a series of ten defense sector reform goals. Each goal is addressed in a separate chapter that marries conceptual clarity and practicality, *describing what needs to be accomplished, identifying a place to start, and providing guidance for how to implement reform*. Each goal is defined in terms of key concepts. For example, what is democratic control and why is it essential for an effective defense sector? What do the terms "democratic accountability," "separation of powers," and "monopoly of force" mean in practice? To bridge the divide between theory and practice, *each goal chapter explains how to select defense sector reform activities and offers real-life examples*. The final chapter addresses some of the thorniest challenges practitioners encounter in designing and implementing defense sector reform. The chapter guides the practitioner through a series of six questions that lead from the design to the implementation of programs and that offer insights into broader security assistance efforts.

With other states increasingly using direct foreign investment and security assistance to further their own foreign policy agendas, the United States and its allies must become more effective and nimble in improving their security assistance approach. This *Practitioner's Guide for Defense Sector Reform* makes an important and very timely contribution to that effort, delivering pragmatic solutions to bridge the divide between theory, policy, and reality. Strategic Capacity Group's extensive field experience and the authors' own expertise and dedication have produced an outstanding contribution to the practitioner's toolbox.

Andrea Dew, PhD
Board Member, Strategic Capacity Group

Acknowledgments

Developing a practitioner's guide to defense sector reform would have been an impossible task without the insight and assistance of countless individuals who assisted our efforts from the perspective of both headquarters and the field.

First and foremost, we would like to thank the Smith Richardson Foundation for its generous support. We are particularly indebted to Marin Strmecki and Nadia Schadlow for their guidance as we developed the proposal for this guide and conducted our research. As we neared completion of the project, Chris Griffin supported our efforts to adjust our research amid the many complications presented by a global pandemic. This guide would not have been possible without their generous and ongoing support of our work.

Snezana Vuksa-Coffman, the deputy chief of the Security Sector Reform Unit at the United Nations Department of Peacekeeping Operations, regularly engaged with us on the United Nations' priorities related to defense sector reform and hosted meetings with her team. Michael McNerney, associate director of the International Security and Defense Policy Center at the RAND Corporation, helpfully guided the selection of some early defense sector reform case studies that informed the development of our ten defense sector reform goal chapters. International practitioners assembled at a conference hosted by the Peace Research Institute Oslo provided invaluable feedback on the form and structure of the goal chapters, and their requests for us to share "the checklist" in a usable format for practitioners substantially influenced the final shape of the guide and its concluding chapter. We are also grateful to staff from the U.S. Department of Defense Office of the Secretary of Defense for Policy and from the Defense Institutional Reform Initiative, including their team leads, for some of our case study countries. Our efforts were strengthened by their thoughtful contributions and their critiques of our early research design.

Over the course of our research, we met with and interviewed hundreds of individuals. In Mali, officials in the Ministry of Defense and the National Parliament, at the United Nations Multidimensional Integrated Stabilization Mission in Mali, and those with defense sector reform mandates at various embassies in Bamako graciously devoted their time and attention to our many questions and spoke candidly about the challenges of reform. We are also thankful for the information many members of the press and international nongovernmental organizations provided to help us better frame our understanding of the reform context. In Georgia, Maka Petriashvili, deputy director of the Human Resource and Professional Development Department at the Ministry of

Defense, and Shalva Dzidziguri, research fellow at the Georgian Center for Security and Development, joined our research team. They helped us to conduct numerous interviews with active and retired government officials in the Ministry of Defense, members of the General Staff, the NATO institution building team, members of parliament, and various civil society organizations engaged in defense sector reform activities. In Tunisia, our research team engaged with the Ministry of National Defense and the Presidency, various parliamentary committees, and numerous civil society groups and experts. In Colombia, our research team included Dr. Rocío del Pilar Pachón Pinzón, director of the Army Center for Strategic Studies, and Nathalie Pabón Ayala, researcher at the Army Legal Department, who assisted us to conduct interviews with Ministry of National Defense officials and members of parliament and the judiciary. Although we do not name all the individuals we interviewed, we are no less grateful for their invaluable contributions to our research and to this guide.

We are also tremendously grateful to Dr. Alejandra Bolanos, who posed—and answered—some of the most difficult questions prompted by our research. Dr. Michael Sullivan provided invaluable research for our Iraq examples and willingly read and critiqued every chapter. Nigel Quinney, our exceptional editor, made the guide usable and readable. Our team at Strategic Capacity Group, including Stephen Allen, Matthew Herbert, Jeffrey Hunter, Oumar Konipo, Harsha Sirur, Alessandra Testa, and Andrea Walther-Puri, contributed many other country examples, assisted with our research and travel, helped prepare the manuscript for publication, and supported our efforts to bring this guide from concept to finished product.

Writing a book-length manuscript is no easy endeavor. Doing so during a global pandemic is even more challenging, and we are particularly grateful for the flexibility and generosity of those who willingly responded to our queries from afar. This guide would not have been possible without their generous and timely support.

ABOUT THE AUTHORS

Querine Hanlon is the founding President of Strategic Capacity Group (SCG). As a professor and dean in the U.S. Professional Military Education system before creating SCG, Dr. Hanlon led the redesign and accreditation of strategic-level education programs for the U.S. military as it grappled with the institutional requirements and needs of the post-9/11 security environment. She designed a post-9/11–focused security studies curriculum for the newest degree-granting college at National Defense University (NDU) and created NDU's first satellite campus at the John F. Kennedy Special Warfare Center and School in Fort Bragg, North Carolina. Dr. Hanlon has also conceptualized and created curricula for partner institutions. At the United States Institute of Peace (USIP), Dr. Hanlon created innovative regional security sector programming in response to the Arab Spring that was recognized by the U.S. Department of State as a one-of-a-kind standard for future programming. Dr. Hanlon has considerable expertise in security sector reform. She has served on security sector assessment trips to Tunisia, Algeria, and Morocco for the U.S. government and has advised government and international organizations on how to design security assistance programming to deliver strategic and sustainable results. She is the author of numerous books and articles, including *Prioritizing Security Sector Reform: A New U.S. Approach* (Washington, DC: USIP Press, 2016). Dr. Hanlon holds a PhD and a master of arts in law and diplomacy from the Fletcher School and a bachelor of science in foreign service from Georgetown University.

Alexandra Kerr is Deputy Director of Programs at Strategic Capacity Group, overseeing SCG's capacity building activities worldwide. Ms. Kerr is a security sector reform expert who has worked in international relations and defense both domestically and internationally for more than a decade. She has considerable experience advising, developing programming for, and working on the ground with security institutions and governments on defense institution building, border security capacity, and transnational organized crime. Prior to joining SCG, Ms. Kerr was a fellow at the National Defense University, where she led research on security sector reform and institutional capacity building in the defense sector, including leading an innovative initiative on U.S. defense institution building policy. Ms. Kerr is the editor and an author of *Effective, Legitimate, Secure: Insights for Defense Institution Building*, a book published in 2017 in collaboration with the Office of the Deputy Assistant Secretary of Defense for Security Cooperation. Previously, Ms. Kerr was Assistant Director of the International Institutions and Global Governance Program at the Council on Foreign Relations, and before that she held research positions at the Center for Humanitarian Dialogue in Geneva and the United Nations in Rome. Ms.

Kerr is a PhD Candidate at the University of St. Andrews in Scotland and holds master's degrees in international conflict from the Department of War Studies in King's College London and international relations from the University of St. Andrews.

About SCG

Strategic Capacity Group (SCG) is an international nonprofit dedicated to building strategic security sector capacity worldwide. Dr. Querine Hanlon founded SCG in 2013 to provide an alternative to the prevailing approach that prioritizes the provision of tools over the building of capacity. SCG's solution is to build strategic capacity to achieve a lasting and measurable change in security provision worldwide. It is based on the premise that donor and recipient efforts are sustainable only when assistance is designed and delivered as part of a strategic capacity approach.

SCG's strategic capacity building approach focuses on enhancing institutional accountability and effectiveness, and strengthening human capacity to design, implement, and lead change. Capacity building is a strategic function. It occurs when knowledge is embedded in processes and institutions that can be shared and disseminated over time and replicated by the recipient even after assistance ends. When assistance is provided through a strategic capacity building approach, it enhances the likelihood of impactful and sustainable results.

SCG helps donor countries to enhance the sustainability and impact of their assistance approaches and recipient countries to deliver security appropriately, effectively, accountably, and in accordance with the rule of law. Through this approach, SCG helps build human and institutional capacity worldwide.

Goals and Principles for Defense Sector Reform

Goal 1: Democratic Control

1.1   Democratic control is exercised over the entirety of the defense sector.

1.2   The prerogatives of the defense sector are clearly enumerated, ideally in the constitution or at least in statute.

1.3   The defense sector is accountable to the population or their elected representatives.

Goal 2: Civilian Control

2.1   A clearly defined and publicly available chain of command establishes civilian authority over the ministry of defense and defense sector forces.

2.2   Civilians participate jointly in the defense sector with uniformed forces.

2.3   Mechanisms promote wider societal oversight of the defense sector.

Goal 3: Legislative and Judicial Oversight

3.1   The defense sector is subject to and complies with regular and transparent oversight by the legislature and the judiciary.

3.2   Information required to exercise judicial and legislative oversight is readily available.

3.3   Members of the legislature have the necessary defense sector expertise to fulfill their oversight functions.

Goal 4: Coordination and Management

4.1   A clearly defined, codified, and resourced mechanism defines how defense and security matters are coordinated with other government ministries.

4.2   Clear policies, procedures, guidelines, and systems coordinate activities, functions, and responsibilities among different offices, departments, and agencies within the defense sector.

Goal 5: Functioning Logistics

5.1   Established systems deliver the five core logistics functions of generation, deployment, supply, sustainment, and transportation.

5.2   The core logistical systems are coordinated and integrated.

GOAL 6: DEFENSE PLANNING

6.1 A planning system directs the efficient identification, coordination, and sequencing of functions and resources to translate strategic goals into operations in the near term, midterm, and long term.

6.2 Defense planners have the right skills to create defense plans and to adapt those plans according to changing conditions.

GOAL 7: FINANCIAL MANAGEMENT

7.1 There is a system for the planning for, allocating, executing, and accounting for resources expended for the defense sector.

7.2 Clearly articulated rules and procedures direct financial decision-making, acquisitions, procurement, and audits.

7.3 Defense sector budgets and budget processes are transparent and subject to oversight.

GOAL 8: THE RIGHT PEOPLE

8.1 The defense sector plans for an effective and appropriate workforce to staff the military and its supporting functions.

8.2 The defense sector has clear processes and policies for personnel management.

8.3 The defense sector provides for the development of personnel skills through an established system for education, training, and exercises.

GOAL 9: STRATEGY GENERATION

9.1 Strategic thinkers can appraise risk, calculate probability and consequences, mitigate danger, address challenges, and prepare for future threats.

9.2 A formal system (or systems) exists for the development and continuous adaptation and revision of the defense strategy.

GOAL 10: MILITARY EFFECTIVENESS

10.1 Doctrine specifies how the military uses its assets on the battlefield.

10.2 Military organizations have optimal command and coordinating structures for the missions they are tasked with conducting.

10.3 Military forces undergo regular and iterative training—including exercises, war games, and other field-based training—to prepare for the missions they are tasked with conducting.

10.4 Defense sector institutions have a capacity and potential for innovation.

CHAPTER ONE

THE NEED FOR
A PRACTITIONER'S GUIDE

Two stories prompted the content and purpose of this guide. The first story concerned a team of experts who had just completed an extensive assessment of a country's defense sector and were providing an initial briefing to the donor embassy team before their flights home. "We had a number of preliminary findings," one of the experts explained, but each finding was "like a Russian nesting doll." Unpacking one challenge revealed others within it. Experienced programmers will likely have a similar story to tell: a broken logistics system reveals corrupt procurement, a "system" dependent on a single individual, and few skilled personnel, each of which is a problem in itself, and no one solution can fix them all. The discussion was framed by a "real sense of urgency," the expert explained, "because security was rapidly deteriorating and there was a great deal of pressure" from the donor capital. The briefing stretched on for more than an hour with no clear way forward, and finally the lead donor staff member interjected, "But can you do something?"

A similar sense of urgency was apparent in another interaction. A programmer had funding that would expire in twelve months, and deployed donor staff were urgently requesting the procurement of material to relieve the country's beleaguered forces. Earlier packages had been rejected for failing to address much-needed institutional strengthening in the recipient country. The programmer was facing growing pressure from the field to "do something" but was at a loss for how to expand the equipment purchase into an institutional strengthening program that could be implemented in a short time frame. I wish there were "something that could help me do this," the programmer told us.

And so, this guide was born. We had already been considering producing an academic work to address two gaps—one between the existing guidance for defense sector reform and the environments where defense sector reform is needed, and the other between the tools for defense sector reform and the push to generate sustainable change in the countries where those tools are deployed. But these two stories, and countless others like them, caused us to rethink our plans and produce instead a practitioner-oriented guide.

Most of the existing literature on defense sector reform and defense institution building is heavily influenced by the successful experiences of Eastern

European countries joining the North Atlantic Treaty Organization (NATO) in the decade or so after the end of the Cold War. Our goal was both to simplify this literature for easy use and reference by the practitioner and to revise it to fit the current context, because the countries where defense sector reform is now most needed—and will most likely be implemented—look very little like Poland or Slovakia. The security challenges are larger and more immediate, important systems are likely broken or missing, and resources are limited. And the powerful incentive for undertaking difficult reforms—membership in NATO and all the benefits it brings—is notably absent. What will compel countries that have serious, and perhaps even existential, challenges, to undertake reform for some future, and possibly uncertain, benefit? Without the proverbial carrot, defense sector reform is much harder to do.

What we learned from talking with defense sector reform practitioners in the United States and Europe, as well as those deployed in countries where such work is currently being conducted, is that determining *what to do*—and more specifically, *where to start*—is often the biggest challenge. This is not surprising, particularly when there is no clear guidance derived from a partnership or alliance such as NATO that dictates programming requirements and timelines. In its place are a range of priorities generated by donor national interests and strategic or operational requirements. These may not align with programming objectives coming from donor embassy or operational staff in the field. Adding further complexity to the task are the recipient defense sector personnel, who may envision a generous assistance package without the attendant reform. Sometimes, donors do provide material or training support to address a near-term, finite gap. The goal in such cases is not to reform how the organization conducts military operations, but merely to help a particular unit at a particular moment in time operate more effectively.

But when the goal is *reform*—real, measurable, and sustainable change—practitioners will need to determine why and for what purpose these efforts are being implemented. If that assistance is meant not just for that one unit at a particular moment in time but to change how the entire force operates, then the challenge is a systemic one. Is the force controlled (democratically or otherwise)? Is it subject to civilian, legislative, and judicial oversight? Is it provisioned and manned appropriately? Does it have an effective strategy and codified doctrine? Can it plan, coordinate, and operate effectively? And if the answer is "No to all of the above," where does the programmer begin?

We propose ten goals for defense sector reform, each of which identifies a place to start and provides guidance for how to implement reform across a range of contexts, from the higher to the lower end of the defense sector capacity spectrum. These ten goals are drawn from leading government and international policy documents, guidance, and handbooks. Chapter 2 provides an overview of the literature that informs this guide. To ensure the ten goals were useful and implementable, we subsequently vetted them with defense sector reform practitioners in both Europe and the United States and used their feedback to further refine the goals. The ten goals are detailed in chapters 3–12, with a chapter devoted to each goal.

Each goal chapter follows a similar structure. First, we define the goal, key terms, and concepts. For example, what is democratic control and why is it essential for an effective defense sector? What do the terms "democratic accountability," "separation of powers," and "monopoly of force" mean in practice? Second, we unpack that goal into its component parts. For example, if the defense sector is democratically controlled, what should the implementer expect to see in terms of structures and functions? Each of these component parts is framed as a principle to guide the selection of defense sector reform activities to achieve that defense sector reform goal. Finally, we provide examples for each principle to show what this could look like in practice in a range of contexts to help the programmer translate the conceptual guidance into action.

The ten goal chapters can be read in sequence, from start to finish, but each chapter provides standalone guidance for programming design and implementation and can be read by itself. If programmers are not sure where to start, chapter 13 offers guidance on how to prioritize and sequence reforms when timelines are compressed or needs are many.

The Goals of Defense Sector Reform

The starting point for any defense sector reform program is to define what that program seeks to achieve. The ten goal chapters are designed to assist practitioners in defining the purpose of the assistance package they are tasked with designing and implementing.

The first goal is **democratic control** of the defense sector. A defense sector is democratically controlled when (1) that control is exercised over the entirety of the defense sector; (2) the elements of that control are clearly laid out, ideally in the constitution or at least in statute; and (3) the defense sector is accountable to the population or their elected representatives. Chapter 3 is a place to look if defense sector reform programming is required to address the presence of nonstatutory forces, prisons, or other "off-book" functions; to "right-size" a bloated, inefficient, or predatory defense sector; to counterbalance an outsized police apparatus; to address continuous "exceptional measures"; or to rebuild a defense sector that has been destroyed by conflict or captured by armed groups. Effective defense sectors should neither dominate the other branches of government nor have the power to exploit legal mechanisms to achieve undemocratic results.

The second goal is **civilian control** over the defense sector. Civilian control exists when (1) a clearly defined and publicly available chain of command establishes civilian authority over defense institutions and forces; (2) civilians participate jointly in the defense sector with uniformed forces; and (3) mechanisms promote wider societal oversight of the defense sector. Chapter 4 is useful for programming when the chain of command does not, ultimately, rest with a civilian or when civilians have little or no access to information or opportunities to participate in the defense sector as personnel or experts. A democratically accountable defense sector cannot function like a "black box," with its inner workings invisible to the public.

The third goal is **legislative and judicial oversight** of the defense sector. Legislative and judicial oversight exists when (1) the defense sector is subject to and complies with regular and transparent oversight by the legislature and the judiciary; (2) information required for oversight is readily available; and (3) legislators and judges have the necessary expertise to fulfill their oversight functions. Chapter 5 offers guidance for defense sectors where such oversight is largely pro forma or absent; where defense sector officials refuse to attend legislative hearings, ignore summons for testimony, or refuse to share budgets, planned procurements, or other information required by law; or where the judiciary is subject to intimidation or corruption. Legislative and judicial oversight cannot be evaded or serve like a rubber stamp if it is to be effective.

The fourth goal is **coordination and management** *between* the defense sector and other government ministries and *within and across* the defense sector. Interministerial and intraministerial coordination exists when (1) clearly defined mechanisms define how defense and security matters are coordinated with other government ministries; and (2) clear policies, procedures, guidelines, and systems ensure activities, functions, and responsibilities among different offices, departments, and agencies within the defense sector are coordinated. Chapter 6 provides guidance for programming when a defense sector struggles to respond to crises or if unnecessary duplication of responses waste resources. Defense ministries must coordinate with the ministries of foreign affairs and finance to fund operations and align those operations with foreign policy goals. Within the defense sector, intraministerial management and coordination are crucial for translating strategic vision into effective day-to-day operations and for allocating resources.

The fifth goal is **functioning logistics**. Effective logistics exist when a defense sector can (1) fulfill the five core logistics functions and (2) coordinate and integrate those functions. These five core functions include generating people, materiel, and services; deploying personnel and materiel (and evacuating them for the purposes of maintenance, reconstitution, and medical care); warehousing, storing, and maintaining the materiel; sustaining the force until the mission is achieved; and transporting materiel and personnel into, throughout, and out of a theater of operations. Chapter 7 provides programming guidance for defense sectors that struggle to provision and sustain their forces and to move much-needed equipment from the warehouse to the front lines. No amount of equipment or technical support will make a defense sector effective if it cannot feed, equip, or provision its forces.

The sixth goal is generating effective processes, mechanisms, and systems to conduct strategic and operational **defense planning**. Effective defense planning exists when (1) a planning system directs the efficient identification, coordination, and sequencing of functions and resources to translate strategic goals into operations in the near term, midterm, and long term; and (2) defense planners have the right skills to create measurable, achievable, and timely defense plans and adapt those plans according to changing conditions. Chapter 8 provides useful guidance for countries that present a long wish list of equipment and training needs but cannot make a case for why it is needed or how it will be used, or for countries that have the necessary resources but no ability to forecast or plan what is needed when and where. Defense planning is a critical component of any effective defense sector; if plans do not accurately capture the defense sector's implementation capacity, the national strategy may not come to fruition. In the absence of effective defense planning, a defense establishment is likely to use resources inefficiently and put lives at unnecessary risk.

The seventh goal is **financial management**. A defense sector has an effective and transparent financial management system when (1) a formal system, or coordinated procedures, direct the planning, allocation, execution, and accounting for resources expended by the defense sector; (2) clearly articulated rules and procedures direct decision-making, acquisitions, procurement, and audits; and (3) budgets and budget processes are transparent and subject to oversight. Chapter 9 offers programmatic guidance where defense sector funds are mostly off-budget, set aside for "government capture industries" (i.e., industries owned by the government and whose profits benefit the government instead of the public), or lost to corruption, or where financial processes cannot sustain operations or other core defense sector functions. Even the most poorly resourced defense sector can do more with those resources when appropriate procedures and processes are in place for directing and tracking their use.

The eighth goal is **the right people**—a simplified term for what is variously called human resources management, personnel management, human capital, or skills-based management. A defense sector has the right people when it (1) can plan for an effective and appropriate workforce; (2) has clear processes and policies for personnel management; and (3) develops personnel skills through education, training, and exercises. Chapter 10 provides guidance for defense sectors that struggle to recruit or retain enough forces for its missions; have undergone a conflict or democratic transition and feature largely inexperienced or newly appointed personnel; or have assignment or promotion practices that reward sectarian affiliations, political loyalty, or corruption rather than military competence. People are at the core of almost any defense sector reform function because they ultimately are responsible for implementing and carrying forward the change. As such, the "right people" may not be the place to begin, but this goal will likely guide programming for any defense sector reform initiative.

The ninth goal is **strategy generation**. A defense sector can effectively generate strategy that links a country's defense objectives to the military and other resources at its disposal when (1) it has strategic thinkers—defined as those individuals who can appraise risk and calculate probability and consequences, mitigate danger, address challenges, and prepare for future threats; and (2) a formal system exists for its development and continuous adaptation and revision. Chapter 11 provides a good starting point for contexts where defense strategies do not exist, are outdated, or collect dust on a shelf because they cannot be translated into operations. Strategies are only effective if the words on paper can be translated into action and continuously updated as needs and conditions change.

The tenth and final goal is **military effectiveness**. A defense sector is militarily effective when (1) doctrine specifies how the military uses its assets on the battlefield; (2) military organizations have optimal command and coordinating structures for their missions; (3) military forces undergo regular and iterative training to prepare for their missions; and (4) defense sector institutions have a capacity and potential for innovation. Chapter 12 is a good starting point for defense sectors that feature absent or outdated doctrine or struggle to translate doctrine into operations; that have high casualty rates for the operational context stemming from poor or absent predeployment training; or that include forces that violate rules of engagement or military codes of conduct stemming from poor command and control. More advanced defense sectors may benefit from steps to remove hurdles to innovation. Put simply, military effectiveness is the "ability of a military force to successfully prosecute a variety of operations against a country's adversaries."[1] As such, it is the sine qua non of defense sector reform.

Context Matters

Most guidelines for foreign assistance programming—whether focused more narrowly on the defense sector or more broadly on social, government, or economic development—feature the admonition that "context matters." No two countries will feature the same gaps or needs or have the same historical legacies, environments, resources, or other conditions that frame what can be done and how to go about doing it. Nonetheless, there are contexts that will be similar, and much can be learned from like contexts.

To translate the conceptual guidance in each of the goal chapters into useful examples to assist defense sector reform practitioners, we generated some criteria for selecting appropriate country examples. First, we sought to select countries that have varying levels of democratic accountability—from more established democracies such as Colombia, where civilian authority is both codified and largely practiced, to countries that have recently transitioned to democratic rule such as Tunisia and Georgia, countries that may finally be emerging from conflict such as Libya, and countries that face serious, and potentially existential, security threats such as Mali. In so doing, we sought at least a modicum of geographic diversity. Some countries, such as Iraq, satisfied more than one of these criteria. A few, such as Iraq and Georgia, have been recipients of many years of defense sector reform assistance, whereas others, such as Tunisia, are relatively new entrants.

We also wanted useful and varied examples from across the defense sector capacity spectrum—countries at one end that feature higher levels of baseline capacity, countries in the middle near the higher and lower ends of the spectrum that have stronger capacity in some areas but gaps in others, and countries at the lower end of the spectrum that have limited capacity and many needs. We chose this spectrum because this is the likely range of countries in which practitioners will be tasked to "do something." Finally, to ensure that there is sufficient depth to our examples—none of which are developed as full case studies—we selected countries in which the authors and Strategic Capacity Group have firsthand experience so that we would not be reliant on secondary sources.

The risk in this approach is that for every example we offer, the reader may well object, "Yes, but that will not work in . . ." Each example we offer will likely have as many as fifty or more counterexamples. This cannot be avoided. There is simply no way to satisfy every reader by including every possible example and still retain the purpose of this guide—a succinct and usable manual written for practitioners. But even if we could add ten or twenty more examples, the problem would be the same. There is no one solution that will work for every context. And every context is different enough that there will always be important exceptions when one country's context is compared with another.

And so, we urge the reader to treat the examples as exactly that—examples that are included to illustrate what a defense sector goal or principle could look like when implemented across a range of contexts. Our purpose here is not to provide *the* answer, but rather to offer some suggestions that might spark an idea or an approach that could work, particularly if the context is similar. As much as it would simplify the programmer's task, there is no one-size-fits-all solution and no definitive way to achieve any of these goals in all contexts. This is why the context matters and why programmers will need to devote effort to understanding that context before determining what programming might achieve the goals we developed to guide this effort.

Programming Pitfalls

We close this introductory chapter with a few guidelines to consider as users read this manual. If, after reading this chapter, the practitioner is still uncertain about where to begin, we suggest turning to chapter 13. It opens with some guidance about how to determine where to start and about sequencing if there are many urgent and important needs. If a programmer has been tasked to "do something" and timelines are compressed, chapter 13 may also help narrow some of the available options, including pointing the reader to the appropriate goal chapters for more detailed guidance. Regardless of where the reader starts to use this manual, it is important to keep a few things in mind at the outset, thereby ensuring that some programming pitfalls can be avoided in the latter design and implementation stages.

First, it is best to give some thought to what success looks like at the outset— and not when programming is already underway and evidence needs to be produced to sustain the program beyond the initial funding or design cycle. A well-designed program should be able to demonstrate that work is being done; who is doing the work; who claims ownership of the work; any evidence of formal adoption; and, most importantly from the standpoint of sustainability, whether the work has become iterative, routine, or ongoing.

Second, programmers will likely initiate some mapping, assessment, or process streaming in the design or beginning stages of implementation.[2] Defense sectors feature many different systems—for logistics, procurement, or strategy generation, for example—and these must be understood. Formal systems are much easier to map because there is documentation, but informal systems are easy to overlook and much harder to map and understand. If it appears that there is no system, there probably is an informal one. The complete absence of a system is rare. Programmers should spend some time, or plan for this time in their program design, to ensure they understand the system that exists. Not doing so will likely waste time and resources, producing activities that may have little impact because they fail to address real gaps or needs. You cannot fix a system if you do not understand how it works or why it is broken.

Third, the programmer will be tempted to launch activities that generate fast and concrete evidence of results—and these are often best portrayed as numbers: numbers of forces trained, number of weapons or other equipment supplied, and so forth. But these number-generating activities are no substitute for real reform. If real reform is the intended outcome, programmers will need to ensure that activities generate institutional capacity so that the assistance they provide can be sustained by the defense sector when the program ends. Similarly, when timelines are compressed and needs are urgent, it is often easier to use what already exists rather than designing something from scratch. What already exists, or is "off the shelf" and ready to implement, is often drawn from what the donor is already using and tends to be more technology-intensive. High-tech solutions tend to generate interest and headlines, but they are less likely to be implementable, particularly in countries at the lower end of the defense sector capacity spectrum. Sometimes low- or no-tech solutions have a much better chance of being sustained after donor assistance ends.

Finally, one of the biggest hurdles to defense sector reform is generating the political will to start and the political commitment to see it through. Programmers should anticipate that reform will engender opposition. Sometimes, resistance manifests as slow-rolling activities (i.e., resisting change by appearing to embrace it but actually implementing reform so slowly that little or no change occurs), because change is hard and generates more work. But more serious and potentially reform-ending resistance may be generated by elements within the government or the defense sector, which see reform as threatening the systems by which they rose to power or maintain their influence. Programmers may be deeply committed to reform and make generous resources available to support it, but none of these efforts will result in more than superficial change or perhaps temporary improvements if the political will is absent among recipient defense sector institution personnel and their political leadership.

Defense sector reform is not something that can be imposed. It must be valued, owned, and driven by the recipient political or defense sector leadership to be successful and sustained. We cannot do it for them, nor can we want the reform more than they do.

CHAPTER TWO

ADAPTING EXISTING GUIDANCE FOR FUTURE DEFENSE SECTOR REFORM

Defense sector reform concerns the transformation, restructuring, or creation of the various components involved in a country's external defense. The term has been defined by the United Nations as "a nationally-owned process intended to reconcile, reform, transform, restructure, reengineer, enhance or develop an effective, efficient, accountable and affordable defence sector which operates without discrimination, with full respect of human rights and, under extraordinary and constitutionally defined circumstances, in support of the establishment, maintenance and upholding of law and order."[1] The defense sector stands apart in the broader security apparatus because of its essential role in "fulfilling the social contract: defending sovereign borders and territories of the state, ensuring the security and prosperity of the citizens therein, protecting the interests and values of the state abroad, and maintaining national and regional stability."[2] As a subcomponent of security sector reform (SSR), defense sector reform takes similar approaches to reform but concentrates exclusively on the apparatuses of defense.[3] According to a study published in 2009 by the Stimson Center, those apparatuses include

> the ministries which develop, manage, and implement defence policy (typically the defence and interior ministries), the bodies charged with oversight of these ministries and their implementation bodies (typically legislative oversight bodies and ministerial internal oversight structures), and the operational actors charged with guaranteeing a country's national security. Operational actors include the regular armed forces of the state (army, navy, coast guard, marines/marine infantry, and air forces); state-sponsored paramilitary forces (gendarmerie or equivalent, and border security forces); customs, and immigrations services; intelligence services; and other organizations that defend the state and its people.[4]

The subcomponent of defense sector reform that relates to the institutional components of the defense sector (often the systems within the defense ministry that are necessary for the broader defense sector to function effectively) is called defense institution building (DIB). DIB efforts seek to help "partner-nation defense institutions to establish or re-orient their policies and structures to make their defense sector more transparent, accountable, effective, affordable, and responsive to civilian control."[5] DIB is based on the premise that defense institutions are the foundation of professional defense establishments; without functioning institutional capacity, other defense reforms cannot be effectively assimilated or sustained.[6]

Although Western countries have been providing defense-related assistance to partner countries for decades, surprisingly little concrete guidance has been developed to delineate the specific aims that defense sector reform ultimately seeks to achieve. Nor does much guidance exist regarding the tools available to defense sector reform practitioners and programmers to achieve these goals across the vastly different contexts encountered in their work. There is also scant guidance on how sequenced assistance activities should be deployed to ensure each institutionalized change lays the foundation for further future development toward the end point of full functionality and efficacy in each area of the defense sector. Defining the specific goals and the principles of defense sector reform need not, however, require starting from scratch.

The ten goals for defense sector reform presented in this guide are derived from a review of policy documents, assessments, and studies that capture lessons learned from defense-focused SSR efforts, as well as interviews with practitioners, programmers, and implementers in the field. Among the many documents reviewed during this study, several useful sources focus (or contain substantive sections that focus) specifically on reform of the defense sector. From these sources, four core documents were selected to frame the basis of this study. Their selection was driven by the frequency with which practitioners cited these four as the documents that guided their country, agency, or organization's policies and programming on defense sector reform. These four documents are

- the Organization for Economic Co-operation and Development's *OECD DAC Handbook on Security System Reform*;
- the United Nations Department of Peacekeeping Operations' "Defence Sector Reform Policy";
- the U.S. Department of Defense's "Directive 5205.82 on Defense Institution Building"; and
- NATO's "Partnership Action Plan on Defence Institution Building (PAP-DIB)."

The following section provides a brief overview of each of these core documents to demonstrate what the existing guidance includes and where this guidance can be expanded and adapted to help practitioners of defense sector reform design and conduct programming across the range of twenty-first-century environments.

OECD DAC Handbook on Security System Reform

The *OECD DAC Handbook on Security System Reform* (hereafter, "OECD DAC Handbook") was developed in 2007 and updated in 2008 by the Development Assistance Committee (DAC) of the Organisation for Economic Co-operation and Development (OECD).[7] This robust compendium on SSR was designed to guide the OECD's 37 member democracies to "close the gap between policy and practice" by "operationali[zing] SSR guidelines" to ensure donor-supported SSR programs are "effective and sustainable."[8] The handbook provides readers with chapters that focus on SSR both conceptually and in practice through nearly 40 SSR case studies.

Although the OECD's focus is traditionally on economic development, the handbook is part of the organization's acknowledgement that a "democratically run, accountable and efficient security system helps reduce the risk of conflict, thus creating an enabling environment for development." Increasing the ability of partners to meet the range of security challenges they face "in a manner consistent with democratic norms, and sound principles of governance and the rule of law" is the overarching objective of OECD SSR assistance. The four SSR principles identified in the handbook can also be broadly applied to defense sector reform: (1) establishment of effective governance, oversight, and accountability in the security system; (2) improved delivery of security and justice services; (3) development of local leadership and ownership of the reform process; and (4) sustainability of justice and security service delivery.[9]

Although the handbook primarily provides insights on SSR, chapter 7 is broken down by sector, including a section dedicated to defense reform.[10] This section acknowledges that although the defense sector "plays a central role in the protection of the state's sovereignty by defending the state against external aggression and internal rebellion," the armed forces can also be a significant "source of insecurity and human rights violations," and thus reform of the defense sector cannot be overlooked.[11] This brief section covers broad entry points for defense sector reform engagements such as transitions from authoritarian rule to democracy, issues for programming and sequencing, some lessons learned, and notes on common challenges.[12]

UN Defence Sector Reform Policy

The UN "Defence Sector Reform Policy" was created in 2011 by the SSR Unit of the Office of Rule of Law and Security Institutions in the UN Department of Peacekeeping Operations (DPKO).[13] The aim of this policy is to guide UN

DPKO staff "to support national DSR [defense sector reform] efforts" and "ensure that DSR efforts contribute to SSR more broadly."[14] It describes the ultimate goal of defense sector reform as follows:

> The goal of the United Nations engagement in DSR is to support national efforts to enhance the effectiveness, efficiency, accountability and affordability of the defence sector and its components, in order to contribute to sustainable peace, security, good governance and development for the State and its peoples without discrimination and with full respect for human rights and the rule of law, and in accordance with national and international norms, laws and nation-specific agreements.[15]

The policy also defines defense sector reform as a subset of security sector reform and something to be pursued in support of SSR.[16]

Although the document largely focuses on defense sector reform within the context of UN peacekeeping operations, the policy identifies ten specific principles that defense sector reform supported by the United Nations should abide by. Of those ten, the following five are particularly relevant to this study:

- DSR shall consider and be sensitive to existing security and defense institutions, concepts, approaches, and cultures.
- DSR shall respect and ensure the commitment of the defense sector and its oversight bodies to national and international norms, laws, and nation-specific agreements.
- DSR shall adhere to the basic principles of transparency, accountability, efficiency, effectiveness, and affordability, while respecting the host nation's right to confidentiality in certain issues pertaining to national security and defense.
- DSR shall aim to strengthen trust and confidence between the State and people within its jurisdiction, defense sector components and other security sector actors, with a view to enhancing the legitimacy of the defense sector.
- DSR shall focus on the development of sufficient national governance, management, institutional, resource (human, material, and financial) and technical capacities and capabilities in the strategic, operational, and tactical dimensions of a national defense sector and its contribution to the overall peacebuilding strategies.[17]

U.S. Department of Defense Directive 5205.82 on Defense Institution Building

Following years of security assistance investments in Afghanistan and Iraq, a growing recognition emerged in the United States that institutional capacity building is the "missing piece" in the U.S. government's security assistance toolkit. This has prompted a renewed focus on incorporating defense institution building (DIB) into the more traditional avenues of assistance such as providing training and equipment to partner nation defense forces. For example, before the addition of DIB, providing a weapons system to a foreign country often meant that the system would be useful to deter an immediate threat but would become defunct shortly thereafter due to the inability of the host country to maintain the system. DIB programming was added to such security assistance to ensure the country would, for instance, have access to the fuel to run the system and the ability and funding to purchase it, access to the parts to fix the system, access to engineers trained to fix and maintain the system, and so on. In the U.S. context, DIB programming aims to make security assistance for foreign defense sectors more effective and sustainable by "building" these countries' institutional capacity, rather than focusing solely on delivering assistance for short-term aims.

The key policy document guiding this renewed focus on DIB is the January 2016 U.S. Department of Defense "Directive 5205.82 Defense Institution Building" (hereafter, "DIB Directive").[18] This document "establishes policy, assigns responsibilities, and provides direction regarding the management, planning, and conduct of DIB by [the Department of Defense]."[19] The directive defines DIB as "security cooperation activities that empower partner nation defense institutions to establish or re-orient their policies and structures to make their defense sector more transparent, accountable, effective, affordable, and responsive to civilian control."[20] According to the directive, adequate defense institutions are those that are "effective, accountable, transparent and responsive to national political systems"; that can "contribute to regional and international security more effectively"; and that can "improve the sustainability, effect, and value of other U.S. security cooperation activities."[21] Doing so requires implementing seven distinct sets of activities:

1. Establishing, building, improving, reforming, and assessing defense institutions.
2. Aligning the defense sector within government-wide systems and fostering synchronization across government sectors.

3. Incorporating principles of accountability, transparency, participation, inclusiveness, and responsiveness, and establishing regulations, procedures, and processes that define their implementation.

4. Prescribing the roles, mission, functions, and relationships within the defense sector.

5. Enhancing the professionalism of defense personnel.

6. Creating or improving the principal functions and duties of effective defense institutions.

7. Promoting institutional operability with allied and coalition forces and institutions.[22]

NATO Partnership Action Plan on Defence Institution Building

NATO's "Partnership Action Plan on Defence Institution Building" (hereafter "PAP-DIB") was first launched by NATO member states at NATO's 2004 Istanbul Head of State Summit.[23] The PAP-DIB was developed with the understanding that "effective and efficient state defence institutions under civilian and democratic control are fundamental to stability in the Euro-Atlantic area and for international security cooperation," but that the reform of defense institutions is "often a long and difficult process."[24] The PAP-DIB was developed to provide support to NATO partners in reforming their defense institutions by "defining common objectives for partnership work in this area, encouraging exchange of relevant experience, and helping tailor and focus bilateral defence and security assistance programmes."[25]

In addition to the original PAP-DIB, which is a relatively short policy document outlining ten objectives for defense institution building as well as a few NATO mechanisms to be employed in achieving these objectives, a compilation of complementary guidelines, entitled *Defence Institution Building: A Sourcebook in Support of the Partnership Action Plan* (hereafter "PAP-DIB Sourcebook"), was published to provide more detailed guidance.[26] The aim of the PAP-DIB Sourcebook is to provide guidance on DIB, "drawing on established practice in Western Europe but also on recent experience among the new members of NATO" by providing case studies, analysis, and descriptions of various components of democratically governed defense systems.[27] The PAP-DIB Sourcebook devotes chapters to each of the PAP-DIB's ten objectives for building institutional capacity in the defense sector.

The first objective is *to develop effective and transparent arrangements for democratic control of defense activities.* Institutionalizing this objective requires implementing democratic control of defense activities in the constitutional, legal, and administrative frameworks of the state. Specifically, this requires defining the roles, division of authority, and relationships among the executive, legislative, and judicial powers, including checks and balances on each in matters of mobilization and use of force, stationing, budgeting, and legislation.

The second objective is *to promote civilian participation in developing defense and security policy.* Here, the guidance is more specific and includes several "requirements." Institutionalizing this objective requires transparency and involvement of civil society in the creation of defense policy; civilians working in defense institutions, especially in leading positions; and transparency in how defense resources are planned and managed. Also required are public dissemination processes to provide information regarding military activities to the media. Furthermore, to achieve "good governance" of the defense sector requires civilian ministers and deputy ministers, as well as military and civilian experts, working jointly in defense agencies.

The third objective is *to establish effective legislative and judicial oversight of the defense sector.* Institutionalizing legislative oversight requires that parliaments can execute five key functions: (1) passing laws that define and regulate defense institutions and their power; (2) adopting budgetary appropriations; (3) reviewing and approving defense and security documents and deployments abroad; (4) participating in defense procurement and personnel management; and (5) holding the executive accountable either through direct questioning of members of government or through special commissions to investigate complaints by the public. Judicial oversight requires that constitutional courts can evaluate the constitutionality of laws and that the judicial branch judges the lawfulness of military behavior and violations of laws on corruption, notably in defense procurement.

The fourth objective is *to develop arrangements and procedures for matching capabilities with security risks, defense requirements, and valuable resources.* This is essentially "strategy-making," or balancing "ends-ways-means" in U.S. terminology. Institutionalizing this objective involves (1) establishing processes for a comprehensive analysis of a country's security needs; (2) involving the executive and legislative branches in developing and approving a strategy; (3) developing supporting doctrine and a military strategy; (4) determining the tasks of the armed forces, including their size and equipment requirements; and (5) developing an implementation plan, including an evaluation of available resources.

The fifth objective is *to optimize management of defense ministries and other agencies responsible for defense matters.* Here the focus is on enhancing interagency coordination, particularly among the ministry of defense and the ministries of foreign affairs and finance to ensure that ministry programs are adequately funded and foreign policy goals are addressed. This also requires a clear demarcation of the roles of each of the ministries with a security mandate in legislation and in agreements and understandings at the ministerial level. Furthermore, it requires the creation of national procedures for crisis management at the strategic level through the creation of an executive supervising body under the leadership of the prime minister or president.

The sixth objective is to *ensure compliance with internationally accepted norms and practices established in the defense sector.* Based on recognized NATO standards for the defense sector, this objective draws on five categories of international norms and standards captured in the "Partnership for Peace: Framework Document" and the Basic Document of the Euro-Atlantic Partnership Council";[28] the Organization for Security and Cooperation in Europe (OSCE) *Code of Conduct on Politico-Military Aspects of Security;*[29] arms control agreements such as the Ottawa Convention; international arrangements on nonproliferation, export control, and weapons of mass destruction and means of delivery; and international humanitarian law. Ensuring compliance involves developing structures and procedures to enforce existing commitments, training personnel to implement those agreements, and releasing public information on how a country complies with these commitments.

The seventh objective is *effective and transparent personnel structures and practices.* Institutionalizing this objective requires the creation of "sound personnel policies" that are essential for an "efficient fighting force."[30] Specific guidance includes establishing clear recruitment and retention policies and extending the same rights and privileges enjoyed by civilians to members of the military (with a few exceptions, such as the right to strike) and with specific attention to how soldiers are disciplined. Furthermore, the guidance suggests that all rules for dealing with military offences should be approved by the legislature.

The eighth objective is *effective and transparent financial, planning, and resource allocation procedures.* There is specific guidance for how this objective is to be institutionalized. Defense institutions should have the capacity to apply modern and efficient planning, programming, and budgeting procedures, including a budgeting and evaluation system and a defense resource management model. Also required are procedures for auditing and oversight of budgeted funds and for awarding contracts for equipment or services.

The ninth objective is *effective, transparent, and economically viable management of defense spending.* Institutionalizing this objective requires developing a clear set of procedures for linking defense spending with overall state budgets to enhance the predictability of the evolution of the defense budget in the medium and longer term, usually through allocating a percentage of GDP to defense expenditures; developing procedures for when and how to prioritize defense spending over other spending, and establishing programs to contend with the socioeconomic aspects of defense restructuring, such as retraining for dismissed personnel and plans for conversion of military bases.

The final objective is *effective international cooperation and good neighborly relations.* This is a broad objective, and guidance for institutionalizing it is equally broad. Recommended actions include implementing bilateral and multilateral military agreements, joint military exercises and training, and regional and defense security cooperation mechanisms. This final objective is critical to the implementation of the other nine objectives because it can contribute to a more stable security environment. A stable security environment can in turn justify efforts to downsize and reorganize defense establishments and achieve cost-saving efficiencies.

Limitations of the Existing Guidance

The various principles, goals, and objectives found in these four documents provide some guidance for conducting defense sector reform, but fall short of providing a comprehensive vision for what defense sector reform should aim to achieve (i.e., what a functioning defense sector should, at a baseline, look like); what exactly would be found in the defense sector (e.g., systems, institutions, policies, mechanisms, capabilities) if the goals had been achieved; and how to go about enacting changes in contexts where elements of the defense sector are missing or wanting. There are three notable gaps in these documents that limit their applicability to a broad swathe of defense sectors and their usability as guides for programmers and practitioners: *the limited context, the absence of a defined endpoint*, and *the lack of specific and widely applicable guidance.*

Context Limitations

The first notable gap is *context limitations*. Each of these documents was written for a specific audience and with specific political aims. Although they were not intended to broadly cover any possible context, widely varying contexts are often commonplace for those using them to guide their programming. In the case of the DIB Directive, the guidance was written for U.S. practitioners, specifically within the Department of Defense, and largely within the context of the U.S. Congress's growing interest in the wake of the Iraq and Afghanistan wars in investment returns on its main tool of security cooperation, train-and-equip programming. Similarly, NATO's PAP-DIB principles are aimed at reform in NATO partner countries and within the incentivized context of NATO partnership or membership. And although the UN Defense Sector Reform Policy does address the defense sector reform concept most directly, the policy is concentrated on the role of defense reform within the limited context of UN peacekeeping operations and only in postconflict environments.

Yet, little attention has been paid to how applicable, appropriate, and realistic the objectives in these leading guidance documents are for states where defense sector reform is currently a priority and where it is likely to continue being so.[31] Although it is widely accepted that no single reform approach will work for all countries, there must be baseline guidance from which to tailor reform efforts. Whether it is a postconflict country, a nascent or transitioning democracy, or a country still mired in internal violence, the overarching goal of a functioning, effective, legitimate, and stable defense sector will apply. This guide is designed to provide guidance that is applicable and adaptable to any defense sector reform context.

Absence of a Defined End Point

The second gap, *what the end point should be,* is also not well defined. Without a clear end point to guide the donor nation's programming, those designing defense sector reform programs are left to make shots in the dark. For example, the OECD DAC Handbook does not define specific defense sector reform objectives beyond the following: "[Defense] reform has multiple objectives—for example effectiveness, efficiency and democratic oversight—that should reinforce each other." The simplicity of this approach to defining overarching goals is attractive, but it leaves those who are tasked with developing programs or "doing something" with little in the way of a North Star to follow: Effectiveness at doing what? Efficiency in what capacities, systems, or mechanisms? What level of democratic oversight is acceptable in the defense sector and how should it be implemented?

Without an overarching defense sector framework, practitioners have to hope that each activity they deliver that is aimed at improving a certain aspect of the host nation's defense will be complementary to other activities and contribute to an overall reform toward an undefined but "better" version of the existing defense sector. As the NATO PAP-DIB Sourcebook notes, the "main functions of ministries of defense worldwide are comparable."[32] All deal with personnel and equipment, intelligence, relations with other ministries, and medium- and long-term planning processes. They also have to manage the recruitment, training, employment, and sustainment of military forces. They are all subject to parliamentary scrutiny, although the mechanisms and degree of oversight varies. And all have decision-making models that place varying degrees of authority in uniformed and civilian authorities such as a chief of defense staff, a minister of defense, or an elected executive.[33]

This guide aims to provide programmers and practitioners with a framework for defense sector reform that defines what should be found in any functional democratic defense sector in the shape of the ten overarching goals: (1) democratic control, (2) civilian participation, (3) judicial and legislative oversight, (4) coordination and management, (5) logistics, (6) planning, (7) financial management, (8) human resources management, (9) strategy generation, and (10) military effectiveness.

Lack of Specific Guidance

Specific guidance on *how to achieve defense reform* is also lacking. The reality is that programmers are not always defense experts, nor are they given specific political guidance other than "do something" with the money allocated to an issue (e.g., "improve the capacity of the armed forces") in a given region or country. But "improve the capacity of the armed forces" is an extremely broad mandate in terms of designing programming and activities. In the four core documents, too often the guidance provided is similarly too vague and provides inadequate roadmaps for successful implementation. In the absence of clear guidance on what defense sector reform is (in terms of goals and objectives) and how to implement it (in terms of programs that build sectorwide effectiveness), these numerous efforts effectively put the cart before the horse. In other words, there is a fundamental imperative to better define what a functioning defense sector means and what programs belong in the defense sector reform toolkit.

For example, although the U.S. Department of Defense's DIB Directive provides a useful set of objectives, it lacks a clear and detailed roadmap for *how* to go about implementing and achieving the DIB objectives. Consider specific items in the directive and the cases of Libya and Tunisia, two countries with widely different levels of capacity. Libya has no functional ministry of defense. In its absence, there is no institutional framework in which to "incorporate principles," nor are there offices where to "improve functions " and "enhance interoperability," or with which to "align other government sectors." Defense sector reform in Libya would require institution building *from the ground up*. There is no guidance for *how* to do this. In contrast, Tunisia does have a functioning ministry of defense—one of its most accountable institutions—and a highly centralized system of government. But it has no political will to invite a foreign donor to conduct DIB activities aimed at "transparency," "accountability," or "synchronization" across the security sector. In Tunisia, doctrine, defense budgets, and even organizational charts remain classified as "state secrets."[34]

As it stands, guidance for how to conduct defense reform in countries such as Tunisia is still incomplete.

Guidance for how to design and implement defense sector reform programs so that they achieve effective and sustainable improvement in these countries' capacity to manage their defense sectors needs renewed attention. In addition to the ten defense sector reform goals, this practitioner's guide provides examples of what these ten goals may look like in varying contexts to demonstrate how different each context can be and to suggest where possible entry points for reform can be found. Finally, each goal has within it "principles" that offer guidance on what the donor nation can expect to find if the country has adequately achieved each of the ten goals. This will allow program designers to easily identify specific gaps in capacity or institutional structures they can target to initiate functionality-oriented reforms.

A Guide for Practitioners

These four core texts and the broader SSR literature on security reform provide the basis on which the ten goals for defense sector reform have been developed. Earlier drafts of the goals and their subprinciples were vetted both in the United States and in Europe with defense sector reform policymakers, academics, practitioners, and implementers to seek their insights, feedback, and critiques. One insight that emerged from these conversations followed a presentation of the goals to a diverse group of U.S., Canadian, and European defense sector reform experts in Oslo. The reaction from this conference was most pronounced among the defense sector reform *practitioners*, many of whom reached for a copy of the manuscript, even in draft form, to help them design their programming. It became clear that defense sector reform practitioners were eager for guidance, especially on how to translate broad policy mandates (e.g., "improving force effectiveness") into actionable and effective programming.

To respond to this need, this guide presents a framework that can be applied across the spectrum of environments where defense sector reform is likely to be needed with the aim of helping practitioners and policymakers better plan for and implement defense sector reform efforts that achieve sectorwide reform rather than short-term and limited change. The ten goal chapters offer a starting point for considering what a functioning, effective, and legitimate defense sector should comprise and what this would actually look like on the ground.

Each chapter takes into account effectiveness, is derived from the real-world lessons in the likely environments where defense sector reform is a priority, and is generalizable enough to meet future or unforeseen requirements. These goals and their subprinciples aim to define what, at a minimum, any defense sector across the broadest possible array of contexts would need to achieve to be considered functional. They provide programmers and practitioners a series of guideposts to determine if a partner's defense sector needs assistance, and in which key capacities there are gaps. Importantly, when considering a partner with a dizzying array of problems, these goals can help practitioners decide where to start.

CHAPTER THREE

GOAL 1
Democratic Control

One of the leading goals of defense sector reform is the effective democratic control of the *defense sector*. The defense sector can be defined in terms of institutions, such as the ministry of defense and the forces it manages and controls (the military, intelligence services, paramilitary units, and, in some systems, gendarmerie, border guards, and private security organizations); the executive (president, prime minister, or head of state) and its staff; and parliamentary committees of defense, security, and intelligence. It can also be defined according to specific defense sector functions, including command and control, strategy and planning, defense intelligence, defense legislation, budget authorization, military justice, human resources, education and training, media and public relations, acquisition and procurement, provisioning, logistics, planning, and deployment.

Democratic control is founded on the concept of democratic accountability, the notion that those who have responsibility or authority to decide upon and implement defense policy and defense sector activities are accountable to elected representatives or directly to the people.[1] Democratic control is often equated with parliamentary oversight, which is an essential but not the sole mechanism through which the defense sector is held accountable in a democratic polity. Rather, democratic control is exercised through an "optimally balanced distribution of power between different institutions of the state" that are ultimately accountable to the population.[2] Put simply, this means that "no single organization should be either so powerful, or so dominant and influential, that it could endanger democratic processes."[3] In practical terms, this involves clear *divisions of responsibility* for missions and services and clearly defined *accountability obligations*, both of which are reflected in *form* (constitutional, legal, and administrative frameworks of the state) and in *practice* (degree of compliance and transparency of state action).

Democratic control of the defense sector is achieved in two ways. The first way is through the distribution of power *among* the different institutions of the state. Thus, the ministry of defense may formulate defense policy, but that policy requires the approval of the executive, the legislature, or both. The ministry of defense may procure new weapons systems, but the legislature has the authority to approve or deny the funds. The selection of officers for promotion to senior ranks may also be subject to approval of the executive or the legislature. And the judicial sector has the authority to review the constitutionality of executive action, legislation, and the legality of operations and to sanction impunity. But the distribution of power alone is not enough to achieve *democratic* control. Authoritarian systems often feature mechanisms for the distribution of power and even for judicial review. What distinguishes democratic from other forms of control are the mechanisms that hold each of these institutions accountable to the population or their elected representatives. Although the exact mechanisms vary from country to country, their purpose is to limit the powers of the defense sector and to hold it accountable.

The second way to achieve democratic control is through the clear separation of powers *between* defense and security institutions. This separation of powers is critical because both defense and internal security institutions wield the state's *monopoly of force*. The monopoly of force is a foundational concept of the modern, Westphalian state system.[4] Legitimate coercive power is concentrated in the sovereign, and the sovereign remains the unique owner of that prerogative, even if the task itself is delegated.[5] Traditionally, that delegation is divided between defense and internal security forces, with the military employed externally and the police internally. The military's conduct is governed by rules of engagement; the police operate according to a different set of use of force guidelines.[6] The separation of powers is also reflected in the ministries that oversee them, with defense forces reporting to the ministry of defense, and police and other internal security forces to the ministry of interior, (homeland) security, or justice. How this separation is defined varies, often considerably, from country to country, but the key issues are that there is a clear separation; that this separation is enumerated in the constitution or statute; and that exceptions—such as for emergencies and national disasters—are clearly defined in terms of what force may be used, for how long, and under whose authorization. What has become increasingly important, as the functional boundaries between external defense and internal security have blurred, is that a clearly defined operational relationship exists between and among forces whose mandates overlap.

The state's monopoly of force and the separation of powers in how that force is applied also have implications for the country's force structure. Specifically, this means that all forces that operate within the territorial confines of the state (over which its writ legally extends) should be *statutory* forces—that is, they are sanctioned by law and are led, managed, provisioned, trained, and deployed by the state (ministries of defense or interior, the executive authority, and ultimately, the people they serve). Although somewhat more controversial, this category of legitimate providers of security can also include various private security actors provided that the state permits them to wield force but ultimately retains "the sole right to use [or authorize the use of] physical violence."[7] The critical issue here is that the privatization of security is "top down"—it is state sanctioned.[8] Separation is thus "both a concept and a system for the division of labor."[9]

It is not surprising that democratic control of the defense sector is prominently featured among the goals of defense sector reform. The reasons for the prominence of this goal over others in the literature can be attributed in part to democratic peace theory, which posits that democracies are less likely to go to war against other democracies. Promoting democratic control of the security sector is thus viewed as essential to international peace and stability. Nationally, it creates the conditions for good governance, institution building, security, economic development, and prosperity. This is reflected in several international norms and standards developed by international organizations such as the United Nations, which "establish parameters as to how defense policy should be conducted within the growing family of international states."[10]

A second reason for its prominence is that democratic control is a requirement for integration and membership in key organizations, such as the OSCE and NATO. Indeed, for some it is the *sine qua non* of membership."[11] The Council of Europe has expanded democratic control beyond the defense sector to include police, security services, and border guards. The most far-reaching requirements are the OSCE standards, which call for the democratic control of military, paramilitary, and internal security forces as well as intelligence services and the police."[12]

Guiding Principles for the Design and Implementation of Goal 1

Although the concept of democratic control of the defense sector is well defined, it is less clear what this means in practice. Worldwide, there is a great deal of variation in how defense sectors are organized. Even among like polities—for example, the member states of NATO—there are great variations in how democratic control is exercised over the defense sector. For countries undergoing democratic transition, the gaps between the conceptual ideal and the reality on the ground are even wider. And for postconflict countries, control of the defense sector, no matter how exercised, is likely to be more of an aspiration than a reality. Nonetheless, there are three guiding principles that can inform how to achieve the goal of democratic control of the defense sector across a wide range of contexts.

First, democratic control should be exercised over the entirety of the defense sector. Quite simply, this means that all defense sector institutions, not just military forces and the ministry of defense, should be democratically accountable. In much of the literature, there is an implicit assumption that defense sector forces and institutions are functionally part of the defense sector. The challenge is not their inclusion, but their control in accordance with democratic norms.

Yet, in many of the environments where defense sector reform is likely to be a priority, the challenge is one of both inclusion *and* control. The first challenge is to identify the actors and forces that constitute the defense sector, either in terms of their lines of authority or their functions. Often, there are secret forces, shadowy ministry departments and offices, secret prisons, hidden arms depots, or specialized units loyal only to the president that do not appear on any public organizational chart, let alone in any ministerial budget—even the classified ones. Military and police units are assigned to ministries and operational commands according to their loyalty, not their function. These complex mechanisms designed for regime protection rather than effectiveness will have to be identified and unraveled. Ministries will have to be divested of some functions, and units and forces may have to be disbanded.

In postconflict states, this task may be less about mapping *statutory* forces and functions and more about identifying armed nonstate actors and remnants of military, paramilitary, and security forces. And if the security or defense sector has essentially collapsed, the objective here will be to build an effective defense or security sector subject to democratic control and oversight. Although the task seems at first glance to be an easy one—mapping who does what and reports to whom—these organizational mapping exercises go to the heart of a state's power structure.

In postauthoritarian and postconflict settings, this mapping often reveals bloated, inefficient, and sometimes even predatory defense sectors that siphon resources, impede economic development, and threaten the lives and livelihoods of populations. In these contexts, establishing democratic control is thus about constraining the power of the defense sector through various "right-sizing" activities involving manpower, budgets, and mission elimination, with a heavy emphasis on professionalization.[13] Pervasive throughout the literature is the focus on limiting the power of the defense sector so that it neither dominates the other branches of government nor holds the power to exploit legal mechanisms to achieve undemocratic results.

In limiting the defense sector, bloated internal security sectors associated with authoritarian police states may be of equal concern. In some instances, a bloated defense sector may serve as a check on a predatory internal security sector, and activities to right-size defense may have unintended consequences when that check is removed. Limiting the power of the defense sector through a distribution of power among other branches of government not only contributes to the goal of right-sizing the defense sector vis-à-vis other sectors, notably the internal security sector, but also further strengthens the defense sector's accountability. Regardless of context, alongside limitations on the power of the defense sector, there should be a clear enumeration of its prerogatives and checks and balances on those prerogatives, particularly involving use of force, manpower, and budgeting.

Second, the defense sector should be accountable to the population or their elected representatives. Although the mechanisms for democratic accountability and oversight vary, their use should ensure that critical decisions involving policies, personnel, finances, operations, procurement of equipment and weapons systems, and international military cooperation are transparent and subject to review and sanction.[14] Additionally, these oversight functions should include multiple and overlapping internal and external mechanisms.[15] In the defense sector reform literature, there is a great deal of emphasis on the types of mechanisms, and variations among democratic systems, for achieving this integration of multiple and overlapping internal and external mechanisms for control and oversight. The assumption is that some oversight mechanisms may exist but that these are largely pro forma or that transparency and compliance may exist but that these are weak or uneven. Thus, efforts have focused on strengthening or streamlining these processes and enhancing their transparency within the context of existing mechanisms to ensure that these are well documented and that incentives exist to promote wider compliance.

However, in weak or transitioning democratic or postconflict contexts, defense sector accountability may be largely nonexistent, and creating it may require far-reaching reforms, including even changes to the constitution. Capacity for exercising this oversight may also be missing, as in the case of parliamentary committees that have the responsibility to review complex budgets and weapons systems but that lack technical expertise required to do so. Even more fundamental, for both newly elected and long-serving defense officials, the entire concept of oversight may be alien, particularly in contexts where disclosures of budgets, weapons, or manpower numbers were previously prohibited. In well-established democratic systems, the concepts of oversight and accountability are part of the institutional culture, and compliance is the norm. But where these processes are new, establishing effective mechanisms for oversight and accountability requires significant capacity-building efforts aimed at those who must execute the review, as well as those who must comply with these new mechanisms.

Third, the prerogatives of the defense sector should be clearly enumerated, ideally in the constitution or at least in statute. This critical task is applicable to a range of defense sector reform contexts. In more established democracies, it may involve capturing established practices in statutes or through constitutional amendments. New or evolving missions, such as counterterrorism, may also require changes to the enumerated prerogatives of the defense sector. In transitioning or weak democracies, redefining those prerogatives in accordance with democratic principles may be undertaken as new constitutions are developed. In postconflict contexts, those prerogatives may have to be (re)defined from scratch, particularly if defense sector forces played a critical role in the conflict or in postconflict stabilization. What receives particular attention in the defense sector reform literature is the enumeration of special or emergency powers, for which the "constitution must stipulate under what terms the forces may so be used, and under whose authorization."[16]

It is not surprising that exceptional measures are highlighted in the defense sector reform literature, given that these can and have been exploited for undemocratic purposes—and in authoritarian states have often been "prolonged indefinitely as long as the regime lasts."[17] Efforts thus have focused on ensuring that the deployment of armed forces for public order service occurs "only when the resources available to the civil authorities to counter serious internal security threats are no longer adequate" and then only in "subsidiary engagement."[18] In practice, this entails ensuring that (1) the armed forces relieve the police only for ancillary tasks (e.g., traffic control, guarding sensitive sites), and that (2) the armed forces assume public order tasks only when no more police resources are available.[19] And given that the power to declare a state of emergency is "the most fundamental power of government," it should be even more constrained, with specific attention to martial law and the potential suspension of civil liberties. Indeed, such powers should be granted only by the constitution or statutory law and be time-limited.

Applying the Goal 1 Principles in Practice

These three principles provide conceptual guidance for how to achieve the goal of democratic control of the defense sector in a wide range of defense sector reform contexts. Their application in practice—particularly when designing defense sector reform activities or interventions—is a contextually driven exercise. Under each guiding principle, examples of how these principles might generate activities or interventions can guide the defense sector reform programmer to translate them into possible activities in a specific defense sector reform environment.

1. Democratic control is exercised over the entirety of the defense sector.

Establishing democratic control over the entirety of the defense sector—and ensuring that all institutions, forces, and actors with defense roles and functions are included in that defense sector as it is defined in statue or the constitution—is a key component of establishing an effective democratic defense sector. For defense sector reform programmers, the first step for establishing such control is to map the actors and forces that constitute the defense sector *as it currently exists*. This mapping should identify what entities carry out defense sector functions (e.g., ministries and their subordinate departments, other offices and agencies, and operational commands and their forces) and how their lines of authority are defined and enshrined in statute or the constitution. Practitioners can use the results of this initial mapping to identify potential entry points for defense sector reform programming. If there are many gaps or numerous functions outside of democratic control, this mapping may also help programmers make important decisions about sequencing.

In states on the lower end of the defense sector capacity spectrum, there may be functions that are not fully under democratic control because processes are missing or functions are conflated in one powerful office or individual. In some other states, authorities over new functions may not be enshrined in statute, or the existing legal framework may be too weak or contain contradictions that can be exploited to limit democratic control over the defense sector.

The more difficult contexts to map are transitioning authoritarian or post-conflict environments. In contexts such as Iraq and Libya, for example, particular attention should be paid to secret forces, shadowy ministry departments and offices, secret prisons, hidden arms depots, or "off-budget" specialized units loyal to the previous regime.[20] This mapping may involve identifying ministry functions that have to be divested or critical functions that are simply absent. In these more challenging contexts, likely places to begin are rosters of personnel who receive payments on a regular basis (thus revealing "ghost payrolls"), provisioning and procurement processes that exist "off-book," or functions that are simply absent, such as policy and strategy generation, because they were concentrated in one or two individuals at the top.[21] In authoritarian states where the internal security sector has played an outsized role or where the defense sector is viewed as a coup threat rather than a regime protector, defense sector institutions and functions may have been deliberately sidelined, as in the case of Tunisia, and programming may have to focus on rebuilding defense sector capacity.[22] In postconflict contexts, this mapping task may be less about mapping *statutory* forces and functions and more about identifying armed nonstate actors and remnants of military, paramilitary, and security forces that operate outside of formal state control. Understanding the complex mechanisms that were designed for regime protection and shielded from elected oversight is a first step to unraveling them.

For example, the successful conclusion of Libya's peace accords will require rebuilding Libya's defense sector. Mapping this defense sector will reveal that it was historically the weakest of Libya's security institutions under Muammar Gaddafi's regime and has now been effectively divided between Libya's Tripoli-based Government of National Accord (GNA) and the Haftar-led Libyan National Army.[23] Decision-making at the executive and ministerial levels is severely impeded by a lack of centralized executive and ministerial-level authority and oversight, ineffective communication and coordination, the absence of clear chains of command, competition between ministries for resources, and the influence of affiliated armed groups over which the ministry has little if any effective control.[24] Past efforts to integrate nonstate armed actors have disrupted defense institution capacity, and many groups operating under those institutions' aegis function autonomously and never received proper training. In 2012, the Ministry of Defense established the Libya Shield Forces, which integrated intact revolutionary groups in the armed forces' structure as a substitute for Libya's army. To maintain political and economic leverage, armed groups that joined the ministry refused to submit to the formal chain of command. Most armed groups under the Ministry of Defense continue to maintain autonomy, and the ministry is often bypassed by direct reporting relationships between the GNA's executive branch and armed group commanders.[25] Unclear chains of command, overlapping institutional mandates, and contested legitimacy inhibit accountability and oversight of Libya's defense sector.[26]

This mapping will also reveal that the Libyan context features right-sizing challenges. A bloated public sector, the absence of personnel management practices, and the lack of training impact performance and organizational capacity. The ministry operates several training facilities, but many of these training assets remain open only on paper.[27] Many training centers are also controlled by armed groups.[28] Although Libya's defense sector is bloated and there are large ghost payrolls, the Ministry of Defense itself is effectively a hollow shell, consisting of numerous departments that are believed to exist in name only. Establishing democratic control over the entirety of the defense sector in contexts such as Libya will be a complex task requiring sustained assistance to rebuild effective defense sector functions under democratic control while divesting functions seized by armed groups, to implement DDR activities to address the presence of nonstatutory armed groups, and to reestablish the state's monopoly of the use of force.

Other contexts may be far less complicated or challenging than the example Libya provides. Potential entry points for establishing democratic control over the entirety of the defense sector will vary significantly by context. Functions that are absent may require generating new institutional processes, building human capacity to implement and sustain them, and enshrining these functions in statute. Weak control will require stronger statutory and constitutional protections. Functions outside of democratic control will need to be demobilized and divested. In postconflict countries that feature nonstatutory forces, establishing democratic control will likely require disarmament, demobilization, and reintegration (DDR) programs to address the presence of armed groups alongside defense sector reform activities to (re)establish national armed forces; a ministry structure to manage, train, and provision them; and a command structure to oversee their operations. Regardless of the scale or complexity of the challenge, knowing what constitutes the defense sector on paper and in practice, what exists outside its control, and what functions are missing is the first step to developing programmatic interventions to strengthen or reestablish democratic control of the sector in its entirety.

2. The prerogatives of the defense sector are clearly enumerated, ideally in the constitution or at least in statute.

Once the defense sector as it currently exists, in terms of lines of authority and functions, has been mapped, activities or interventions can be designed to strengthen or establish defense sector prerogatives through practice, in statute, or through the drafting or revision of the constitution. In transitioning authoritarian states where the functions of defense and internal security overlapped, defining defense sector prerogatives may also require a new or revised division of labor and, where roles do overlap—as, for example, during a state of emergency or natural disaster—that those roles are clearly defined and time-limited.

For example, in Georgia, after a new constitution was adopted in 1995, additional laws were passed to address "multiple contradictory and ambiguous clauses which often became a source of sharp competition between state institutions, hindering the institutionalization of defense reforms."[29] Additional laws were also passed to strengthen democratic control of the defense sector. The 1997 Law of Georgia on the Defense of Georgia designated the Parliament of Georgia as the body that defines state defense policy, approves military doctrine, passes defense laws and the budget, and approves force levels; the president as the supreme commander in chief of the armed forces who submits military doctrine to the parliament for approval; and the Ministry of Defense as the agency that implements defense policy and oversees the combat readiness of the armed forces. The subsequent 1998 Law on the Status of Military Servicemen further enhanced parliamentary oversight of the military, particularly for military appointments.[30] Almost a decade later, further revisions to the prerogatives of the defense sector were captured in the 2006 Law of Georgia on Defense Planning.[31] In contexts such as Georgia, defense sector reform activities may be designed to support the crafting of legislation to address contradictory or ambiguous clauses, whereas in countries where new constitutions are being drafted, assistance can be provided to the drafting process to ensure that democratic control of the defense sector is clearly enshrined.

In some contexts, particularly in postauthoritarian states, the challenge will be one of right-sizing or otherwise limiting the power of the defense sector or of the executive. Such legal reforms may be required even after a new constitution has been adopted. Here, the case of Georgia is again illustrative. Despite amending the constitution and passing additional legislation, other steps were taken to limit the president's power. In 2013, to limit the power of the newly elected president, Giorgi Margvelashvili, the Parliament of Georgia passed amendments that assigned certain national security and crisis management functions to a newly created Council for State Security and Crisis Management.[32] Similarly, to enhance democratic oversight, Georgia's parliament established a three-member Trust Group (a chair and representatives of the ruling party and opposition party) to monitor confidential military and security programs.[33]

In countries such as Mali and Tunisia, violence has generated the exercise of emergency powers, which have often been renewed year after year. Defense sector reform activities in such contexts may involve either addressing loopholes in the law governing the declaration of emergency powers or focusing on the drivers that generate government reliance on emergency powers to provide security. If the latter is the case, likely activities will fall under many of the other goals of defense sector reform beyond *Goal 1: Democratic Control*, such as *Goal 10: Military Effectiveness* (including command and control and training of forces) and *Goal 4: Management and Coordination* to address the cross-ministerial coordination for shared mandates, such as border security. Across all contexts, the priority here will be for programmers to identify where and how assistance can be provided to clearly define and capture the prerogatives of the defense sector. The initial mapping of the defense sector *as it exists* will help the programmer to determine where and how to focus that assistance.

3. The defense sector is accountable to the population or their elected representatives.

Although the mechanisms for democratic accountability and oversight vary, their outcome should ensure that critical decisions involving policies, personnel, finances, operations, procurement of equipment and weapons systems, and international military cooperation are transparent and subject to review and sanction by the population or their elected representatives.[34] The initial mapping will identify where there are critical gaps and where subsequent programming can start with a more detailed assessment to understand specifically what limits that accountability.

In most defense sector reform contexts, an accountable defense sector will be achieved through defense sector reform activities to advise and guide the development of new or revised laws, constitutional amendments, or in the case of postconflict or postauthoritarian transitions, new democratic constitutions. Where those already exist, activities may focus on enhancing transparency to further undergird defense sector accountability by establishing additional oversight bodies or creating enhanced functions or processes. For example, in Colombia, the Ministry of National Defense published a plan in 2015 to "ensure integrity and prevent corruption," and created an Army Transparency Standards Board to enhance accountability and transparency of its anticorruption efforts following the peace agreement with the FARC (or, to give it its full name, the Fuerzas Armadas Revolucionarias de Colombia, the Revolutionary Armed Forces of Colombia).[35] Often, mechanisms and oversight bodies exist but cannot exercise their authority because of a lack of compliance, accountability, or even staff expertise. The constitution may establish legislative authority over the defense budget, but that authority may be difficult to execute if the defense sector can evade legislative summons to present the budget for approval and testify as to its expenditures or otherwise undermine the oversight process. Similarly, the constitution may stipulate that the size and rank structure of the armed forces is determined by the legislature, but this clause will have little meaning if the defense sector is not required to submit records for review, ignores legislative authority when assigning personnel to senior appointments, or presents budgets for legislative review that lack sufficient detail to verify the size of the force against what is being expended on personnel. There is no one-size-fits-all solution for how this can best be achieved; programmers will need to spend time understanding the specific context, and the initial mapping will be a place to start.

In many of the contexts where defense sector reform will be a priority, democratic control may exist in statute but not yet in practice. The first elected president of Georgia never fully gained control over the leadership of the Georgian Armed Forces, which generated a "chaotic pattern of civil-military relations" and contributed to a military coup in December 1991, led by the chief of the National Guard and the leader of a nonstatutory military group declared illegal by the new president.[36] The first president's successor secured control over the military in 1992 but not through democratic institutional mechanisms. Instead, he relied on personal influence and the "old method of divide and rule, never going after all the warlord challengers at one time."[37] Indeed, it was not until 1995, when Georgia's new constitution was adopted, that democratic control of the defense sector was enshrined. The constitution clearly named the president as the supreme commander of the armed forces and gave parliament the authority to approve force levels. A year later, a national security council was established to advise the president on the organization of the country's defense and its military forces.[38] As noted above, initial limitations on parliament's authority to revise the budget proposed by the Ministry of Defense were subsequently addressed in a law passed in 1997, further enhancing the defense sector's accountability. In a context like Georgia, defense sector reform activities will likely need to focus on constitutional amendments and laws to better define how oversight of that sector is exercised.

In some defense sector reform contexts, mechanisms for elected or popular oversight of the defense sector may exist but are not followed in practice— usually because the capacity to do so is absent. For example, Tunisia's new constitution clearly establishes elected oversight of the defense sector, but parliamentary subcommittees assigned this oversight function struggle to deliver it because they lack the detailed knowledge of what the defense sector does and how it functions. In transitioning states where the defense sector previously functioned as a black box, this lack of basic knowledge is often a major impediment to the exercise of appropriate oversight. At the same time, ministry officials who have never had to present to the public their budgets, or weapons acquisition plans, or any other information (whether via their websites, reports made available to the press, or testimony or hearings before parliament) will need to learn what oversight means and what they are required to share, at what level of detail, and how best to share this information with people who are not defense sector experts. Here, too, a good entry point for defense sector programming is basic capacity-building activities for ministry officials and their oversight counterparts. Additional activities to help enhance transparency can also increase the defense sector's accountability to the population or their elected representatives.

CHAPTER FOUR

GOAL 2
Civilian Control

A second goal of defense sector reform and defense institution building is to promote civilian control. In the defense sector reform literature, this goal is defined as requiring civilian control *over* the defense sector and civilian participation *in* the defense sector (as a means of achieving that control).

Civilian *control* is achieved, in part, through measures to ensure democratic control of the defense sector—constitutional or statutory mechanisms that establish a "chain of *civil* command," usually in hierarchical order, that includes the president of the state, the prime minister, the ministers, and those in charge of each of the defense sector forces and institutions.[1] In some polities, the prime minister may be the most senior civilian in charge of the foreign, security, and defense policies of a country, with relevant ministers subordinate to him or her. In either case, the chain of command ensures that civilian ministers are superior to the heads of the armed forces, police, secret services, and other security forces and are accountable to the president/prime minister and to parliament. In practice, this means that a "democratically elected top executive civilian leadership" is responsible for assessing threats and that security and defense forces are "acting, each separately but coordinated, in accordance with that threat assessment."[2]

Civilian *participation,* which is an explicit requirement in the second PAP-DIB objective, is achieved through several mechanisms, including the presence of civilians working in defense institutions, especially in leading positions. Specific mechanisms for achieving this objective include appropriate recruitment, training, and retention (career development, pay, promotion) policies and programs. But there are also broader requirements for achieving civilian participation that reflect SSR principles, including requirements for *transparency* and *good governance.* Here, mechanisms include civil society participation in how defense resources are planned, managed, and employed through public dissemination processes involving the media and through public education initiatives involving journalists, academic experts, and nongovernmental organizations (NGOs). Additional guidance emphasizes not just civilian participation, but civilians and military personnel working *jointly* in defense agencies.[3]

The concept of civilian control of and participation in the defense sector draws on civil-military relations theory. According to Samuel Huntington, a "complex balancing of power and attitudes among civilian and military groups" is necessary to achieve a "system of military relations which will maximize military security at the least sacrifice of other social values."[4] According to Huntington, "Nations which develop a properly balanced pattern of civil-military relations have a great advantage in the search for security. . . . Nations which fail to develop a balanced pattern . . . squander their resources and run uncalculated risks."[5]

Civilian control of and participation in the defense sector is essential to defense sector reform for several critical reasons. First, when power legitimately changes, it is not only civilian control, but also civilian participation within the defense sector that helps ensure that the armed forces "dutifully serve their new political masters."[6] In many of the countries where defense sector reform is likely to be a priority, military forces are trained and equipped for *regime* rather than *national* security. Introducing, or increasing, civilian participation—in the form of both political appointees and career civil servants—in the defense sector can critically shift the balance in favor of those loyal to newly elected democratic leaders and newly institutionalized democratic processes. A greater preponderance of civilians who embrace the principles and practices of transparency and good governance can also open institutions to public scrutiny and influence a gradual shift in institutional culture that embraces more transparent decision-making and citizen engagement.

Second, civilian participation can also promote greater interagency cooperation and coordination between the defense sector and other security ministries. In more established democratic polities, civil servants often move among various government agencies, and senior ministry positions are held by political appointees who report to the country's elected leadership or senior political decision-making bodies such as a privy council or security council. Civil servants' ability to move within and across government ministries ensures that both expertise and knowledge are more widely diffused, reducing the likelihood that the defense ministry can operate like a black box. In transitioning authoritarian states, the introduction of a civilian workforce into previously closed ministries is particularly important when processes and decisions are based on personal relationships in the absence of established institutional processes. Under these conditions, the introduction of civilian staff loyal to the new democratic regime can, over time, erode explicitly designed coup-proofing mechanisms and institutional cultures that prohibited coordination and information sharing.

Third, when civilian participation is broadly defined to include a country's domestic security community of analysts, academics, journalists, interest groups, and other civil society organizations, then there will be a wider societal oversight of the defense sector. Such broad societal participation and oversight is viewed as a "hallmark of good governance in advanced democracies."[7]

Fourth, civilian participation can further widen and deepen the relationship between the military and society at large, which can in turn influence the extent to which the military is well integrated into society. Without such integration, the danger is that the military effectively exists as a "state within a state."[8] Civilian participation can also serve to depoliticize the military's role in society and, depending on how civilian participation and decision-making are structured, minimize political interference in professional military matters.

Indicators of integration include patterns of recruitment as well as popular attitudes toward the armed forces. There is, however, an important caveat. Interestingly, the literature highlights the abolition of conscription as a means of promoting defense sector modernization and institution building through the increased professionalization of the armed forces and the commensurate reduction of national defense budgets. Although the literature acknowledges that ending conscription can decrease the level of integration between the military and the wider society, "the enhanced professionalism of the soldier to some extent compensates the negative effect."[9] Although this may be true in the context of established democracies in Western Europe, it is not clear that the same argument applies elsewhere. In his seminal study of the role of armies in transitioning states, Zoltan Barany cites the cautionary example of Iraq:

> The dismissal of the Iraqi army created an alarming security and public safety vacuum; produced a large pool of trained, armed, humiliated and desperate men for whom joining the anti-American insurgency became a logical choice; and destroyed the only national institution in a deeply divided society, an institution that could have actively participated in postwar reconstruction.[10]

Although the end of conscription does not mean the abolition of the armed forces, the loss of employment opportunities for young males can have significant social repercussions in transitioning democratic and postconflict states that could in turn imperil a peaceful transition to democratic rule. In multiethnic societies, ending conscription could also create conditions that enable the "capture" of the armed forces by a privileged (ethnic) minority.

Fifth, most but not all Western ministries of defense employ a large number of civilians to work alongside military officers because civilians offer specialized skills, such as administration, management, and finance, that military professionals may lack.[11] However, promoting rapid civilianization can produce unintended consequences, as in the case of the Central and Eastern European countries, which, in reaction to Western pressure in the wake of the Cold War, rapidly produced "civilians" in their defense ministries. Most of these personnel were former military officers who "retired in place." This situation occurred partly because of the dearth of civilian expertise available in post-Communist countries, but in part it reflected the residual belief in the primacy of the military in defense matters.[12] True civilianization of the defense sector may require a more gradual implementation in countries where defense sectors are almost exclusively staffed by uniformed personnel to ensure that integration of civilians—and their participation in the defense sector—is achieved in both form and substance. This will require "maximiz[ing] the particular skills of both civilians and the military, professional or retired, and ensur[ing] that they complement and reinforce each other."[13]

Given that "there are no hard and fast rules for the number of civil servants in the ministry of defense,"[14] the exact distribution of uniformed and civilian personnel will be context-specific and likely determined by a range of factors, including budget (military personnel are usually more expensive to employ in terms of salaries than civilian personnel, and they retire earlier), skills gaps, the degree to which human resources processes are able to manage civilian hiring, the pool of available and willing civilian talent from which to recruit, and the availability of civilian academic programs that focus on security. Functions where civil servants are viewed to be indispensable are the directorates dealing with general policy (as distinct from the general staff), financial control, and, increasingly, procurement and personnel because "acquisition procedures and labor conditions are approaching practices in the civilian sector."[15] And, because they are less likely to rotate frequently, civilians can serve as expert professional staff who retain institutional memory and inform decision-making by rotating uniformed or political officials.

Guiding Principles for the Design and Implementation of Goal 2

Although the concept of civilian participation in and control of the defense sector is widely recognized as a key goal for any defense sector reform effort, guidance for how to do so across a range of polities is less well defined. Many Western democracies feature well-established and powerful civil servant corps in their ministries of defense, whereas transitioning authoritarian states may have defense sectors that are exclusively staffed by uniformed personnel. Three guiding principles inform how to achieve the goal of civilian control over and participation in the defense sector across a wide range of contexts.

First, a clearly defined and publicly available chain of command should establish civilian authority over the ministry of defense and defense sector forces. This means that the armed forces and all other defense sector forces (gendarmerie, border forces, secret services, and other specialized units, as well as any private security forces operating under a ministry of defense contract) are under civilian control and that this chain of command is enshrined in constitutional or statutory mechanisms that are publicly available and establish clear lines of authority, including the authorities of both civilian and military leaders for the generation and effective use of military force.

Second, civilians should participate jointly in the defense sector with uniformed forces. How civilian participation—let alone working jointly—is achieved will vary depending on the context. In some instances, as in the case of most established democracies, civilians and uniformed personnel working jointly means that civilians work alongside uniformed personnel throughout the defense sector, hold leadership positions, and constitute a core professional staff that supports more frequently rotating uniformed officers. Where civilian participation numbers are less robust, joint participation may still be achieved through the types of positions held by civilians—for example, positions critical to decision-making (e.g., senior ministerial staff) or requiring different types of expertise (e.g., weapons systems procurement planners). However, in transitioning or postconflict contexts, where civilian numbers are low or nonexistent, achieving this goal may require interim steps to avoid the risks generated by rapid civilianization or integration. For example, civilian participation may be limited initially to the inclusion of external civilian expertise for discrete or high-profile tasks (e.g., strategy development) or to undertake specific functions that are reform priorities (e.g., personnel or procurement). Over time, as the participation of civilians gradually contributes to a shift in institutional cultures, civilians may be introduced in additional and more sensitive functions throughout the ministry.

Establishing benchmarks for civilian participation in the defense sector should be designed for and sensitive to the local context. Likely factors could include tangible measures such as the size of the budget, specific skills gaps, the ability of the organization's personnel department to absorb a new or larger class of employee, or the size of the pool of available and willing civilian talent from which to recruit. Other less tangible but equally important considerations include the institutional culture of the target defense organization(s), the context in which these reforms are being implemented and under whose direction, and the degree to which existing personnel are willing to accept potential shifts in their power and prerogatives. Both categories of factors will significantly impact the design of any program to achieve greater civilian participation in the defense sector. Ultimately, the goal should be to enhance the long-term security of the government and its citizens by reducing the risk that the armed forces function as a state within a state.

Third, mechanisms should be designed to promote wider societal oversight of the defense sector. These mechanisms will vary greatly by country and will be influenced by the number and type of analysts, academics, journalists, and interest groups, and other civil society organizations that could potentially play a societal oversight role. Whereas such communities in Western democracies tend to be large and well established with high levels of knowledge, access, and expertise, in many of the countries where defense sector reform is likely to be a priority, these groups are often nascent, tiny, or lacking in expertise or professional capacity because the defense sector was previously closed or otherwise inaccessible to nonuniformed personnel. When public study of the defense sector becomes possible—for example, following democratic transition—defense institutions struggle to find experts in their countries with real or deep expertise to inform their new planning, policymaking, or strategy-making activities; defense institutions also have difficulty finding journalists with the requisite knowledge to report accurately and responsibly on defense sector issues. In the interim, as these communities slowly grow in size and expertise, false or misleading reporting or poorly constructed guidance can embolden former regime holdovers who continue to view any "civilian" participation with deep distrust. In these contexts, mechanisms should aim both to build the capacity of these communities and to create mechanisms that will lead to greater societal oversight.

Applying the Goal 2 Principles in Practice

These three principles provide conceptual guidance for how to achieve the goal of civilian control over and participation in the defense sector in a wide range of defense sector reform contexts. Their application in practice—particularly when designing defense sector reform activities or interventions—is a contextually driven exercise. Under each guiding principle, some examples of how these principles might generate activities or interventions are provided to help guide the practitioner to translate them into possible activities in a specific defense sector reform environment.

1. A clearly defined and publicly available chain of command establishes civilian authority over the ministry of defense and defense sector forces.

Mapping what constitutes the defense sector of the country slated for possible defense sector reform programming and its chain of command is a good starting point because it will quickly reveal whether that information is publicly available and, on further investigation, whether it is complete. This mapping should include identifying all the component institutions and forces within the defense sector and their chain of command, including any private security forces with a defense mandate. For example, the Law of Georgia on the Defense of Georgia defines the Georgian defense sector as the Ministry of Defense (Civilian Office), the general staff, the armed forces, and six other legal entities, including the Defense Institution Building School, the National Defense Academy, the Cyber Security Bureau, the Military Hospital, the Cadet Lyceum, and the Military Scientific Technical Center Delta.[16] The Ministry of Defense has responsibility for implementing national defense management and civilian control over the Georgian Armed Forces, and the parliament must approve the use of force.[17]

Such information is less readily available in Tunisia, which began its democratic transition more recently than Georgia.[18] The Tunisian Armed Forces consist of the Army, Navy, and Air Force, and report to a civilian minister of defense. However, the Tunisian military, unlike the armed forces of other Middle East and North African countries, never played a political role, even prior to President Zine El Abidine Ben Ali's ouster in 2011. Tunisia's armed forces did not receive any special compensation or material advantages for their service to the state.[19] "In minimizing and sidelining the military, Ben Ali created a professional, apolitical institution that was little known—and little feared. . . . Having expressed a preference for the barracks over politics, the armed forces emerged [from the Arab Spring] as one of the only respected institutions of the ancien regime."[20] Civilian control of the military is also enshrined in Tunisia's new constitution. Tunisia's most senior military officer reports to a civilian minister of national defense, who serves as a member of the prime minister's council and reports to an elected head of government.

Despite clearly defined chains of command codified in statute, both examples reveal that additional reforms may be required to fully realize civilian control in practice. According to Georgia's Law on Defense, the general staff has little influence and no authority over the appointment of military officers, or their promotion to higher ranks, or for the selection of flag officer candidates for review and approval by the parliament, which raises the risk that appointments and promotions are subject to "political interference" or that officers are appointed to positions not for their suitability but their political loyalty.[21] In Tunisia, the reform challenges reside in the embedded residual culture and practices of the old regime—most notably in the form of an intelligence cell within the armed forces that has a largely internal function, namely, monitoring and reporting on the activities of its officers.[22] Uniformed officers within the Ministry of National Defense also have little to no role in contributing to the development of defense strategy. "We are not free to voice our ideas or have to be very careful when we do so," one officer explained. "It is much better now than in the old system, but we cannot yet shape policy, make strategy, or engage with them [other government officials, politicians, or citizens] outside, although we hope to do so."[23] The challenge here is posed not by an absent chain of command but by very tight control of the uniformed chain of command.

To achieve the goal of civilian control and participation, defense sector reform activities need to be designed to ensure that a civilian chain of command exists both in form and in function. Achieving this goal might require improving processes for the selection and promotion of officers by developing mechanisms that incorporate the military chain of command in the selection process, as is often the case in more established democracies. Other activities might focus on improving civilian control by reducing mechanisms, including undemocratic levers, that limit the professional functions of uniformed officers on which effective chains of command rest. Still other useful interventions might include training, perhaps in concert with activities envisioned under other defense sector reform goals, to improve how military and civilian counterparts function within the defense establishment, starting with a clear understanding of the delineation of their respective roles in line with the codified chain of command.

In environments where many defense sector reform activities have already taken place, such as Georgia, or in more recently transitioning states, such as Tunisia, defense sector reform activities may focus on longer-term as well as ongoing efforts to shape the institutional cultures of defense sector institutions so as to undergird civilian control or to increase the transparency of those structures. Where the initial mapping reveals limited available public information, enhancing citizen access to information about the defense sector and civilian oversight of it may be a place to start.

In other contexts, a clearly defined chain of command establishing civilian authority over the defense sector may exist and be codified in statute, but the absence of effective democratic checks and balances or of oversight mechanisms limit abrogation of the chain of command in practice. In Iraq, for example, a civilian chain of command was enshrined in Iraq's 2005 constitution,[24] but real civilian control rested with the prime minister only, who oversaw the creation of a sectarian and inefficient military institution with no effective checks or balances.[25]

After the fall of Mosul in 2014, military leaders blamed Iraq's civilian leadership for the rapid collapse of the four Iraqi divisions defending the city, accusing them of failing to provide adequate guidance and funding. Civilian leaders in turn accused military leaders of

> massive corruption, high rates of absenteeism, poor training standards, and a complete lack of unit cohesion. Fearing a political challenge from the officer corps, in the eight years preceding the defeat in Mosul, Prime Minster Nouri al-Maliki deliberately exerted tight control over the military, intervening in nearly all personnel and equipment decisions.[26]

Maliki unilaterally dismissed his most seasoned officers, replacing them with candidates selected for their loyalty or tribal affiliations, many of whom had been previously removed for corruption and human rights violations. The result was a hollow force commanded by individuals with little combat experience that was quickly overrun by Islamic State of Iraq and Syria (ISIS) fighters.

Maliki also sidelined the Ministry of Defense and the Security and Defense Committee of the Iraqi parliament and used the Office of the Commander in Chief to bypass checks on civilian control, appointing an ally to head the office and promoting him to the rank of general. Once he took control over the portfolios of the ministers of defense and interior, the Office of the Commander in Chief became the executive body for the entire Iraqi security system.[27] Those who tried to oppose Maliki's consolidation of power were either punished or made irrelevant.[28] The minister of defense, Abdul Qadir Obeidi, tried to depoliticize the ministry and the armed forces and was barred from running in the 2010 elections.[29] Other politicians who tried to resist the consolidation of power in the prime minister's office were similarly sidelined; oversight mechanisms were either weak or nonexistent; and the deteriorating security situation in and around Baghdad made any attempts to prevent this consolidation futile.[30]

In postauthoritarian or postconflict environments, activities aimed at generating civilian control will likely be more foundational, aiming either at *establishing* civilian control or ensuring that a chain of command does more than exist on paper. Such activities will be less technical and more political because they aim at (re)defining power restructures at the heart of the state. Where such changes are not imposed by force—through, for example, the defeat of a powerful faction vying for state control—they are often best addressed through a renegotiation of the state's social contract, activities which more likely fall under *Goal 1: Democratic Control.*

2. Civilians participate jointly in the defense sector with uniformed forces.

In most established democracies, civilians work alongside uniformed personnel throughout the defense sector, hold leadership positions, and constitute a core professional staff that supports more frequently rotating uniformed officers. In countries selected for defense sector reform programming, determining the extent and nature of that civilian participation begins with assessing available information about defense institution staff sizes and their composition. As with mapping the chain of command, this is a good starting point because it will reveal the extent to which information about the defense sector is publicly available, to whom, and at what level of detail. Does a ministry website give only a total count of staff or does it break down their composition between civilian and uniformed staff? Or are staff sizes only found in budget documents released to parliament and thus publicly available, but not easily accessible to the general public or civilian watchdog organizations? Furthermore, are civilian employment opportunities in the defense sector publicized and open to the public? Or does most recruitment and appointment for civilian staff positions target uniformed personnel who have recently retired?

Where civilian participation numbers are less robust, joint participation may still be achieved through the types of positions held by civilians—for example, positions critical to decision-making (e.g., senior ministerial staff) or requiring different types of expertise (e.g., budgeting or procurement). If information about staff composition and numbers is not readily available, or if defense sector institutions are largely or exclusively staffed by uniformed personnel, further investigation may be warranted to determine how civilian participation alongside unformed staff is achieved.

Numbers tell only part of the story. For example, a ministry may have few civilian positions, but they are all placed in key decision-making units or leadership functions. Alternatively, there may be large numbers of civilians, but only in the most junior or supporting positions with little meaningful participation in setting defense policy or informing resource decisions. There may also be designated offices or functions that are largely civilian within a larger uniformed unit. For example, the Georgian Ministry of Defense's Civilian Office is staffed by 450 personnel, 95 percent of whom are civilian, and its mandate is to "support the minister in managing the Georgian Armed Forces [GAF] and developing policy in [the] defense area and monitoring its implementation, managing and allocating resources, planning and conducting procurements, planning and executing defense budgets, implementing international and public relations, conducting internal audits, and inspecting for the GAF."[31] Since 1991, Colombia's Ministry of National Defense has operated under the direction of a civilian appointed by the president and all administrative and financial processes, as well as the identification and compliance with strategic objectives, are under civilian control.[32] Military matters are largely left to uniformed personnel, whereas civilian staff are responsible for overseeing military compliance with strategic goals and objectives.[33]

In states transitioning to democracy, civilian personnel tend to be nonexistent, few in number, or relegated to support functions. In such a context, defense sector interventions may require a series of interim steps to gradually increase civilian participation overall or civilian participation in key decision-making functions. The risk of pushing rapid civilianization may generate "retirements in place" of uniformed personnel, thereby achieving the goal of participation in numbers but not necessarily in effect. Programmatic efforts could seek to civilianize a critical office or function first, perhaps in line with a defense sector reform priority (e.g., improving procurement processes or reducing corruption), or could focus on recruiting civilians for a particular area of expertise in which there is a reservoir of trained civilians in the private sector, such as finance and budgeting. Over time, as the participation of civilians gradually contributes to a shift in institutional cultures, civilians may be introduced in additional, or more sensitive, functions throughout the ministry. In all contexts, greater civilian participation may require developing a cadre of specialized civilian expertise, which may necessitate specialized training or recruitment initiatives. Here, activities may overlap with those to achieve *Goal 8: The Right People*.

3. Mechanisms promote wider societal oversight of the defense sector

Mechanisms that promote wider societal oversight of the defense sector will vary greatly by country and be influenced by the number and type of analysts, academics, journalists, and interest groups and other civil society organizations that could potentially play a societal oversight role. In Georgia, the Civil Council on Defense and Security, the Center for Strategy and Analysis, the Atlantic Council-Georgia, and the Center for Civil-Military Relations are civil society organizations that have been active in security and defense issues since 2005. They frequently participate in roundtable discussions to support defense reform and to develop and revise defense security concepts and strategies.[34] Similarly, in Colombia, several civil society organizations are active in monitoring defense and security policy, and military policy is openly discussed and debated in the media and by civil society organizations.[35] Passed in 2003, Law 850 allows citizens to monitor the performance of any public body through "civilian watchdogs" (*veedurias ciudadanas*), although there has been public reticence to employ this mechanism for the military sector. As a result, civilian society is active in monitoring defense activities, but less engaged in influencing defense and security policy.[36] In 2016, in an effort to enhance transparency, the army created a new Office for the Application of Transparency Norms in the National Army (DANTE) and launched a "Line of Honor #163" phone line and email address for defense officials and members of the public to anonymously report corruption or misconduct by the police, armed forces, or other bodies overseen by the ministry.[37]

In such contexts, defense sector reform activities could focus on enhancing transparency mechanisms or strengthening ways in which civil society can inform the development of defense sector policy and strategy. Likely activities could focus on sponsoring events at defense colleges or think tanks to generate greater interaction among defense sector institutions and civil society counterparts. Even where such mechanisms already exist, useful activities can support greater engagement to enhance transparency and build trust.

Whereas civil society in Western democracies tends to be well established, in many of the countries where defense sector reform is likely to be a priority, civil society is often nascent, tiny, or lacking in expertise or professional capacity because the defense sector was previously closed or otherwise inaccessible to nonuniformed personnel. Tunisia's defense sector has been largely isolated for decades, and very little information was known about the defense sector or its internal workings under the Ben Ali regime.[38] "The topic of security and defense issues," one official explained, "was closed to those outside this system. There were effectively walls between the sector and the general public and civil society organizations before 2011."[39] After the transition, a few civil society organizations with security mandates began to emerge but struggled to gain access. Few had expertise on defense matters, because the topic was largely taboo for researchers in Tunisia, and there was little available funding within Tunisia for such work.[40] One such organization, the Tunisian Center for Global Research, was founded by a group of retired officials and officers to shape the role of the military in the post–Ben Ali period.[41] Within the academic community, few experts had knowledge of defense matters and no private foundations or think tanks funded or sponsored this work. By 2017, some research activities were being conducted, but largely by trusted entities tied to the government or the defense ministry, such as the Military Research Center. Another institution—the Institut Tunisien des Etudes Stratégiques (ITES, the Tunisian Institute for Security Studies)—is tied to the Tunisian presidency and supported a review of security and defense policy to develop a "Tunisia 2025 Strategy" for the government.[42]

As in the case of Tunisia, when public study of the defense sector becomes possible—for example, following a democratic transition—defense institutions struggle to find experts to inform their new planning, policymaking, or strategy-making activities or to find media with the requisite knowledge to report accurately and responsibly on defense sector issues. In such cases, defense sector reform activities could prioritize building the expertise of researchers and other experts through workshops, conferences, or other activities that bring experts together with those from other countries. Exchange programs could also foster a nascent expert community, and small grants could help establish new institutions or think tanks. At the same time, activities may need to focus on building trust between outside experts and defense sector officials, particularly where such relations were previously marked by distrust. In such cases, building expertise among a small, relatively closed group initially may generate trust that can expand over time to a larger community of experts. Other activities could promote the study of political science, defense studies, or security studies at local universities, possibly through student exchanges or visiting faculty positions with international academic or military education institutions, to begin to generate a new generation of academics with an interest in and expertise on defense and security matters. Graduates of such programs could be recruited to work as civilians within the defense sector or in civil society organizations with mandates to study and inform defense sector policy and strategy.

CHAPTER FIVE

GOAL 3
Legislative and Judicial Oversight

A third goal of defense sector reform is to establish legislative and judicial oversight of the defense sector. *Legislative oversight* involves the formal and regular review of defense sector policies and budgetary actions. Although the mechanisms by which this oversight is conducted will vary by polity, legislative oversight requires that parliaments can execute five key functions: (1) passing laws that define and regulate defense institutions and their power; (2) adopting budgetary appropriations and scrutinizing defense sector procurements and outlays; (3) reviewing and approving defense and security policies and deployments abroad; (4) participating in defense sector personnel management; and (5) holding the executive accountable either through direct questioning of members of government or through special commissions to investigate complaints by the public. *Judicial oversight* involves the formal review and evaluation by constitutional courts of the constitutionality of laws governing the defense sector and by judicial branch judges of the lawfulness of military behavior and the violation of laws on corruption, notably in defense procurement.

Legislative and judicial oversight of the defense sector is essential to defense sector reform for several critical reasons. First, the legislature serves to hold governments accountable through actions that range from "advise and consent" to direct authorization. Declarations of war, deployments, and states of emergency may be subject to prior legislative authorization or ex post review. In some parliamentary systems, budgets are submitted annually for review and approval. In others, each line is subject to review and scrutiny before authorization. Legislatures may also review defense sector policies or play an active role in their formulation through hearings and other consultations. Promotions of military officers and civilian appointments to the defense sector may also be subject to legislative review or approval, and the pay and benefits for military personnel may require legislative authorization. Although the form and function of that oversight will vary, effective oversight ensures that citizens, through their representatives in the legislature, can both shape and evaluate the government's defense and security functions, decisions, and actions and ultimately hold the government accountable through elections.

Second, effective oversight—that is, oversight that exists not just in form but also in function—requires a degree of transparency balanced against the requirements of secrecy. Oversight functions may be codified in law, rules, or procedures, but if the defense sector can withhold essential information about defense sector actions, policies, decisions, or expenditures such that these cannot be meaningfully reviewed by legislative or judicial bodies, then oversight becomes little more than a rubber stamp. In many of the countries where defense sector reform is likely to be a priority, weak or newly created legislatures struggle to access the information essential for them to fulfill their oversight functions. For there to be effective oversight, the legislature must have access to information even if that access is granted to some subset of the legislature, such as a specialized committee, due to the sensitive nature of the information being reviewed.

Third, effective oversight requires that members of the legislature have the expertise to fulfill their oversight functions. Modern defense sector policies, budgets, and weapons procurements are marked by a level of complexity that often requires a high degree of specialized knowledge to fully review and understand. In well-established democracies, members of specialized defense and security committees are often supported by large civilian staffs who know the intricacies and past histories of key decisions. Parliamentary research units may provide additional support for members who are responsible for defense sector oversight. Members often have large personal staffs to assist with the collection of information, scheduling of meetings and review sessions, and the provision of essential briefings in advance of formal hearings. All this support ensures that members can access the information needed and develop the expertise to fulfill their oversight functions. In many of the countries where defense sector reform is likely to be a priority, however, legislative bodies lack much of this support, and members who are assigned to defense and security committees have little knowledge about the most basic structure or functions of the defense sector and little opportunity to gain or develop that expertise, further limiting their ability to provide effective oversight, even if the information is available and shared.

Fourth, effective oversight requires the existence of mechanisms—either codified in law or procedure, or otherwise grounded in precedent—that enable the legislature and the judicial branch to exercise periodic oversight. Legislative oversight is effective only if it is exercised and exercised regularly, such that accountability mechanisms in the form of public hearings or budgetary reviews cannot be cancelled or rescheduled by government officials who would rather avoid public scrutiny. Similarly, in the judicial sector, the requirements for judicial review—either of new laws or of cases involving corruption or violations by uniformed or civilian personnel referred to the courts for administrative or criminal sanction—must be observed regardless of the political affiliation of those who author the new laws or the identity, rank, or seniority of named individuals or units in the referred case. In many of the countries where defense sector reform is likely to be a priority, however, oversight may be required by law but not exercised in practice because opportunities to review policies, budgets, or decisions or to hold hearings with key ministry personnel are perfunctory, sporadic, or too infrequent to provide more than a superficial understanding of the defense sector's workings. In the judicial sector, weak adherence to the rule of law may provide opportunities to influence case assignment to friendly judges or to apply political pressure to ensure dismissal. Ad hoc military tribunals or extralegal processes may be wielded effectively to ensure that judicial oversight is effectively meaningless.

Guiding Principles for the Design and Implementation of Goal 3

Although the concept of legislative and judicial oversight of the defense sector is widely recognized as a key goal for any defense sector institution building and reform effort, guidance for how to create such oversight across a range of polities is less well defined. Many established democracies feature robust oversight processes and procedures that are well codified in law or precedent, whereas transitioning authoritarian states may struggle to overcome deeply grounded resistance to public scrutiny of actions and decisions that have long been protected from public view. Three guiding principles inform how to achieve the goal of legislative and judicial oversight of the defense sector across a wide range of contexts.

First, the defense sector should be subject to the regular and transparent oversight of the legislative and judicial branches of government. Although the mechanisms for accountability and oversight vary, the outcome should ensure that critical decisions involving policies, personnel, finances, operations, procurement, and international military cooperation are transparent and subject to formal review.[1] Effective oversight subjects defense sector decisions and actions to review and sanction in the public domain, precluding backroom deals and poorly vetted ventures that generate corruption and high costs in blood and treasure.

The mere existence of oversight mechanisms is not sufficient; the defense sector must also fully comply with codified law, procedure, or precedent requiring periodic oversight by legislative and judicial bodies. This means that the defense sector cannot evade, reschedule, or simply ignore summonses for hearings or official requests for information from designated parliamentary review bodies or from the judiciary. Although the mechanisms whereby compliance is mandated may vary by polity, oversight must be codified to preclude arbitrary or politically motivated deviations from full compliance. Not only must this oversight be required, and failures of compliance sanctioned, but it must also be sufficiently periodic to ensure that the public, through legislative and judicial bodies, can regularly scrutinize and hold the defense sector accountable for its actions and decisions. For judicial oversight to be effective, violations of the law must be uniformly applied and protected from political machinations that enable corrupt or illegal actions to be free from sanction.

In weak or transitioning democratic or postconflict contexts, defense sector oversight may be largely nonexistent, requiring far more substantial reforms, including the introduction of new statutes or even changes to the constitution. Where mechanisms for oversight do exist, reforms may need to focus on the actual exercise of oversight. For example, hearings may be scheduled, but members may receive cursory or no information. In some instances, ministry officials simply ignore an invitation to appear or send junior ministry officials who have no authority or knowledge about the issue on the agenda. In many such countries, the concept of oversight is new, and the transparency it requires is perceived as threatening the power and prerogatives of the defense sector. In these cases, reforms must also help shift the institutional culture to one that at least recognizes, if not yet values, the importance of oversight in a democratic system of government.

Second, effective judicial and legislative oversight requires that the information needed to exercise real oversight is readily available. In well-established democracies, the challenge is usually one of finding the right information among the plethora of reports and documents released by government agencies. Additional information analyzing the impact of proposed legislative or judicial actions, policies, or other major decisions may be available, adding additional context for those entrusted with a formal oversight function. Additionally, lawmakers can schedule hearings and testimony to further delve into the nuances of proposed actions subject to legislative review, and the courts can compel testimony if so required. Often, the challenge is not securing access to information, but having the time and inclination to review it.

In weak or transitioning democratic or postconflict contexts, however, the information required to conduct real oversight is often withheld or simply missing. In transitioning and postauthoritarian states, information may be available within defense sector ministries and agencies, but laws or practice prohibit its release. In some instances, newly appointed officials may wish to release information, but the processes for doing so do not yet exist, or there is no platform—such as a website or publications register—through which to release it. In postconflict contexts, oversight may be hampered by the absence of information because records were destroyed during the conflict or because violence or operational requirements made the collection, collation, publication, and dissemination of information to legislative bodies, to the public, or for judicial review impossible.

In other instances, classification rules have not yet changed such that nearly all information—including even basic information such as unit names and numbers, base locations and purposes, personnel rosters, and budgets—is subject to classification or special handling rules prohibiting its release. Effective oversight requires that even classified information be subject to appropriate and limited review in accordance with clearly defined and delineated secrecy guidelines that are uniformly applied. In transitioning polities where even the day-to-day functions of the defense sector were only recently treated as "classified," ministry officials may resist sharing information, citing secrecy concerns. In many cases, ministries shield themselves from the requirements for consultation by claiming that secrecy trumps the legislature's need to know. In such instances, even the most basic information may be withheld, limiting the authority of the legislature and thus the essential mechanism whereby the government is held accountable.

Third, members of the legislature should have the ability to develop or access the necessary defense sector expertise to fulfill their oversight functions. Established Western democracies often feature robust systems for ensuring that legislatures have the expertise to exercise their oversight functions. Such expertise may be provided in the form of large permanent staffs and parliamentary research units. Committee assignments may be used to ensure continuity of oversight by longer-serving members. Newly elected members may receive orientation sessions at the start of their term to better prepare them for their roles. The larger and more complex the defense sector, the more important these systems are for ensuring that oversight is truly effective. However, in many of the countries where defense sector reform is likely to be a priority, legislative bodies lack much of this support, and members who are assigned to defense and security committees have little knowledge about the most basic structure or functions of the defense sector and little opportunity to gain or develop that expertise, further limiting their ability to provide effective oversight, even if the information is available and shared. Where the concept of oversight is new, establishing effective oversight requires significant capacity building efforts aimed at those who must execute the review, as well as those who must comply with these new mechanisms.

Applying the Goal 3 Principles in Practice

These three principles provide conceptual guidance for how to achieve the goal of legislative and judicial oversight of the defense sector in a wide range of defense sector reform contexts. Their application in practice—particularly when designing defense sector reform activities or interventions—is a contextually driven exercise. Under each guiding principle, some examples of how these principles might generate activities or interventions are provided to help guide the practitioner to translate them into possible activities in a specific defense sector reform environment.

1. The defense sector is subject to and complies with regular and transparent oversight by the legislature and the judiciary.

Mechanisms for accountability and oversight should ensure that critical decisions involving policies, personnel, finances, operations, procurement, and international military cooperation are transparent and subject to formal review.[2] Even well-established democracies may undertake reform of their legislative and judicial oversight processes in response to operational mandates, public pressure, or executive or legislative direction. For example, Colombia has a military criminal justice system that prosecutes military officers accused of violating the law. Law 1407 of 2010 describes the military penal code, containing more than six hundred articles outlining unlawful conduct; type of sentences; precautionary measures for military personnel under investigation or indictment; and other items in the jurisdiction applicable to the military.[3] Colombia undertook reform of its military justice system in response to demands that security forces be held accountable for abuses.[4] Promulgated under Legislative Act No. 2 of 2012, the reform proposed that crimes committed by the security forces would be tried by the military criminal justice system, while crimes against humanity, genocide, forced disappearance, extrajudicial execution, sexual violence, torture, and forced displacement would be taken to the ordinary justice system.[5] In 2013, this law was struck down by the Colombian Constitutional Court, which declared the law to be unconstitutional. In 2015, the president removed the military criminal justice court from the military's jurisdiction and declared it an autonomous entity to ensure greater judicial independence of rulings relating to service personnel in the line of duty.[6] Despite these efforts, the president of the House of Representatives' Second Committee has noted that more reform is warranted: "Military criminal justice has remained static over time and should adapt to the transitional justice clauses of the peace agreements with the FARC."[7]

Effective oversight subjects defense sector decisions and actions to review and sanction in the public domain. Colombia's constitution stipulates the oversight functions of Colombia's legislature. The Second Congressional Committee can summon officials from the Ministry of National Defense and the armed forces to question policy implementation and has the power to approve or reject high-level appointments to the military and the national police, which report to the Ministry of National Defense. Congress's Third and Fourth Committees exercise budgetary oversight and have the power to "scrutinize, amend, and reject . . . the defense budget."[8] However, "off-budget military expenditure, which is not formally authorized within the country's official defense budget . . . is not clearly regulated."[9] Furthermore, the Ministry of National Defense is required to hold public accountability hearings on the use of resources and to allow for public questioning by citizens. Finally, 2011 reforms created the Office of Internal Control, which is staffed predominantly by civilians to enhance its independence, to counter corruption within the Ministry of National Defense.

In contexts such as Colombia, defense sector reform interventions may focus on strengthening oversight mechanisms—by, for example, enhancing the independence of the judiciary—or generating greater compliance with existing policies or mechanisms. In other contexts, particularly where democratic processes have been only recently established, defense sector reform initiatives may be more fundamental, including establishing the most basic oversight mechanisms where previously there were none.

In Georgia, for example, a new constitution was adopted in 1995, four years after the country declared independence. Although the constitution established legislative and judicial oversight of the defense sector, vague language and the absence of compliance meant that little if any oversight was exercised in practice.[10] In the aftermath of the transition, a shortage of food and uniforms, poor living conditions, and the absence of personnel policies and systems led the Georgian Armed Forces to inflate employment figures, generating large numbers of "ghost" soldiers on the payroll to increase its share of the budget.[11] The absence of effective oversight—and the lack of data management systems to reveal inflated personnel numbers—generated growing and largely unchecked corruption. Financial planning was also largely absent. From 1991 until 2003, mismanagement of tax collection coupled with high inflation rates also took a toll on the central budget, and the Ministry of Defense received only half of the budget it had planned for. To make up the difference, the ministry engaged in extracting illegal, extra-budgetary revenues such as selling military equipment and fuel, embezzling public funds, and collecting "unregistered income from farms under its control."[12] Senior ministry officials signed "a large number of dubious contracts with suppliers," which often were owned by their relatives despite national laws prohibiting such activities.[13] Despite recurring allegations of corrupt defense sector procurement, the Parliamentary Trust Group, responsible for scrutinizing classified procurement and services, never recorded an incident of wrongdoing by a ministry official.[14]

For many years, parliament was given only a one- or two-page defense budget, providing little opportunity to investigate expenditures thoroughly.[15] Its first opportunity to review a "program budget" did not come until 2002. Other procedural constraints also limited oversight—for example, only the president could submit the budget to parliament, and once submitted, parliament had only two options: to approve or to reject the budget in its entirety.[16] Legislators responsible for reviewing classified programs received no additional information from their counterparts reviewing open source information. Subsequent legislation did attempt to address some of the oversight deficiencies. In 2013, constitutional amendments reduced the powers of the president and increased the authorities of the government and prime minister over the security sector.[17] Nonetheless,

> preventive parliamentary control is comparatively weak. Parliamentary control in frequent cases acts in response to the activity of the executive and, consequently, it . . . is often conducted quite ineffectively. . . . It is necessary to strengthen control over military procurements and . . . to use the mechanisms of effective accountability in practice regularly; therefore, the further control of the Parliament should also be intensified.[18]

Subsequent reforms in a context such as Georgia's would likely focus on better defining the clear division of responsibility between parliament, the ministry, and the armed forces because existing gaps allow the ministry "to sidestep the parliament."[19] Furthermore, the "lack of institutionalized oversight mechanisms such as regular and frequent inquires [into] how [the ministry] implements the concepts, big projects, and reforms further weakens Parliamentarian control."[20] The prospect of future reform initiatives is made more likely by the widespread recognition—shared by government, international officials, defense civilians, military officers, and civil society organizations—that legislative oversight is weak.[21]

In more recently transitioning authoritarian or postconflict states, defense sector oversight may be largely nonexistent, requiring far more substantial reforms, including the introduction of new laws or even changes to the constitution. In Iraq, for example, one of the key hurdles to effective judicial oversight has been a campaign to intimidate Iraq's judiciary. Programmatic efforts to counter this effort included increased use of personal security details, personal protection, and moving high-profile cases to alternate jurisdictions. In 2010, improvements in judicial security produced a slight increase in the transparency of legal proceedings involving the Iraqi Security Forces (ISF) and their interactions with the civilian population.[22]

Where defense sector oversight mechanisms do exist, reforms may need to focus on the actual exercise of oversight. For example, Tunisia's parliament faces challenges executing its oversight functions. In addition to a lack of expertise and access to information, the Security and Defense Committee, which is entrusted with executing oversight of the defense sector, has no mechanism to enforce compliance and no authority to review the defense budget. Although the committee meets regularly, usually weekly when parliament is in session, agendas frequently change, meetings are often rescheduled, and membership shifts frequently.[23] "Absenteeism is also a real problem," commented one member of parliament, with only half of the committee's 22 members attending a meeting at any given time.[24] The government's budget is reviewed in its entirety by the Finance Committee, but the Security and Defense Committee "has no decision making authority over the [Ministry of National Defense's] budget."[25] According to members of the Ad Hoc Committee on Administrative Reform and Good Governance, the problem is that the "former system issued guidance from the top down. Now we have a parliament with 270 members entrusted with oversight, legislation, and regional prerogatives."[26] Committees can submit questions to government ministries in written form, but ministerial hearings and testimony are not part of established practice. According to one member of the Security and Defense Committee, "we need to compare the nice words we are hearing from senior officials with the money they are spending. We don't do this. . . . Within the security sector, we don't know what they are doing with that money."[27] Although ministry officials and even operational commanders are invited to testify, in some instances they ignore the summons, and in others they appear but provide little information. Committee members also note that there is no public record they can access to verify information they receive. When some members have recommended inviting outside experts to testify, the invitation has been blocked by members who "don't want an outsider."[28]

In weak or transitioning democratic contexts such as Tunisia, designing defense sector reform interventions to enhance judicial and legislative oversight may require translating new authorities into practice. The concept of oversight is new, and legislators may not understand how new regulations can enable them to compel testimony or why such testimony may be essential to their oversight function. Similarly, ministry personnel may be unfamiliar with their obligations as ministry officials or unused to the public discussion of information that was previously closely held or even considered classified. In other instances, they may be willing to provide information but may themselves lack access to it because there is no infrastructure to capture and disseminate reports, budgets, manpower studies, and other key documents. Often, such transitions require fundamental shifts in institutional cultures—and these are not easily addressed by changes in statutes or procedures. More often than not, such shifts will require longer-term and more fundamental reorientations in how defense, judiciary, and legislative officials define their own roles, functions, and purpose in a new democratic system. For example, in Tunisia, there is no orientation program for newly elected members of parliament to learn how the body functions or the requirements of their new positions.[29] Useful places to begin may include professional development or training on how oversight functions and why it matters, possibly combined with an introduction to how parliamentary or judicial oversight works in more established democratic polities through, for example, a site visit, a short-term assignment in an embedded role in a ministry or parliamentary committee, or mentorship relationships with serving ministry or judicial officials or members of parliament.

2. Information required to exercise judicial and legislative oversight is readily available.

A principal hurdle to effective oversight is access to the information required to exercise that oversight. In established democracies, the challenge is usually not one of access, but of finding the right information in a timely manner and scheduling appropriate levels of review. Of course, there may be instances where subjects find ways to evade their oversight obligations or delay their required response, but mechanisms can be used to compel or encourage compliance. Oversight bodies have recourse to a range of actions and sanctions for noncompliance if the information they require is missing or withheld.

Where oversight is less well-established, however, a basic hurdle to effective oversight is access to information. It is also a critical entry point for defense sector reform interventions. The case of Tunisia is illustrative. Tunisia's legislature is a relatively new body, and its specialized committees began operations in 2014. Members of parliament have no permanent staff to help track and review information, schedule meetings with officials, or otherwise support their oversight functions. Only the committees are assigned aides, and their numbers are small. For example, the Security and Defense Committee has only 3 aides for its 22 members.[30] Unlike the robust and senior permanent staff that such committees rely on in more established democracies, these aides are mostly administrative support staff. They do not, according to one member, "have the requisite skills to fulfill their functions."[31] Oversight is also hampered by reporting delays. There can be as much as a two-year lag between a request or event and the release of a report.[32] "The results are then dated and overdue. . . . This is a government-wide problem, but it greatly affects our work."[33]

Classification rules may hamper oversight, particularly in transitioning authoritarian states where even basic information such as unit names and numbers, base locations and purposes, personnel rosters, and budgets is subject to classification or special handling rules prohibiting its release. Even if those rules have changed, practices are often deeply embedded and take time and training to reverse. In Tunisia, members of parliament noted that even when ministry officials meet with special committees, little relevant detail is shared. There is a widely shared sense that oversight functions are being stymied by officials who are uncomfortable with sharing information that was tightly controlled under the regime of Ben Ali.[34] In such instances, defense sector reform interventions may need to focus not just on pushing for more transparency on routine information about ministry functions, personnel, and budget expenditures, but also on training relevant ministry staff on why and how such information should be shared to meet democratic oversight requirements. In other instances, improving or streamlining repository mechanisms for routine reports or information sharing may be helpful—at least with generating greater access to routine information that is not classified but nonetheless closely held—as a first step to improving oversight through better access to information.

3. Members of the legislature have the necessary defense sector expertise to fulfill their oversight functions.

Access to information is one important hurdle to establishing more effective oversight of the defense sector, but effective oversight also requires that those with responsibility for conducting that oversight have the appropriate expertise to fulfill it. Established Western democracies often feature large permanent staffs and parliamentary research units that can help provide the expertise new legislators—or experienced legislators assigned to new committees—may lack. Committee assignments may also be used to ensure continuity of oversight by longer-serving members, and newly elected members may receive orientation sessions at the start of their terms to better prepare them for their roles. However, newer democracies may lack this kind of support, and members who are assigned to defense and security committees may have little knowledge about the most basic structure or functions of the defense sector and little opportunity to gain or develop that expertise, further limiting their ability to provide effective oversight, even if the information is available and shared. For example, Tunisian legislators assigned to the Security and Defense Committee noted that they know little about the sector they oversee,[35] and there are also very few Tunisian experts who can be called on to testify.[36] Committee assignments change frequently so members do not have the opportunity to develop that expertise over time. Retired military members who founded one of Tunisia's only NGOs focused on the defense, security, and intelligence sector similarly highlighted the lack of military expertise of parliamentary members, noting that there have been visits to defense installations to engage members of parliament on the workings of the defense sector.[37]

In Georgia, the challenge is compounded by a lack of interest on the part of the public, which in turn generates little attention from parliament. Ministry officials echo the lack of interest, noting that parliament "does not require hearings about defense plans on particular policies frequently enough."[38] Even procurements "are not controlled thoughtfully."[39] The Parliament Trust Group started controlling classified procurements only after the Ministry of Defense leadership suggested such a legislative initiative in 2013.[40] Members of parliament attribute this lack of interest to low public participation in the political process, which in turn leads to "insufficient accountability from the government or the Parliament."[41] Because of low public interest on defense and security issues the parliament does not feel pressured to hold the ministry accountable.[42] Furthermore, parliamentary members lack expertise in defense issues and do not have the technical knowledge to engage in discussions when ministry officials present reports to parliament. As in Tunisia, absenteeism is also a problem. Parliamentary members often do not attend meetings related to the defense budget and to other important defense sector reform issues.[43]

Programmatic interventions to generate greater expertise on the part of parliament—or on the part of the judiciary when reviewing cases involving the defense sector—can be addressed through professionalization and training interventions. Public apathy is more difficult to address given the scope of the challenge, although greater transparency and media reporting might stimulate more public interest. Programmatic entry points might involve encouraging local civil society institutions, think tanks, and universities to sponsor events and activities to begin heightening knowledge and awareness of the defense sector, which could, over the longer term, begin to address some of the expertise and interest gaps among the legislature and the general public.

In countries where civil society features few organizations or experts who work on defense matters or have expertise in the sector, simple programmatic interventions could help to build such expertise. For example, donors could support the establishment of organizations, perhaps as subsidiaries of more established NGOs, as some European donors have done in Tunisia, or academics in Tunisia's nascent think tank community or university professors could be invited to attend Tunisian or international workshops and conferences focused on the defense sector or to participate in research projects with a similar focus. Building a community of experts in a country's oversight bodies, and supporting their work by generating more local expertise, can be important steps in generating the access to the information and know-how required to exercise the oversight prerogatives that may already exist in the constitution or statute.

CHAPTER SIX

GOAL 4
Coordination and Management

The fourth goal of defense sector reform is the creation of mechanisms for coordination *between the defense sector and other government ministries* (interministerial coordination) and for management of functions, resources, and decisions *within and across the defense sector* (intraministerial management). The complexity of the modern state, the international environment, and defense and security operations have made inter- and intraministerial coordination an imperative. Defense ministries must coordinate with the ministries of foreign affairs and finance to fund operations and align those with foreign policy goals. Within the defense sector, intraministerial management and coordination are crucial for translating strategic vision into effective day-to-day operations and for allocating resources.[1]

Defense ministries often share mandates for counterterrorism, border security, and intelligence with other ministries and their subagencies, requiring close coordination of planning and operations from the strategic to the tactical levels. Most modern states have developed executive-level coordination bodies, such as national security councils, to manage the normal, day-to-day flow of information and decision-making at the strategic level. Their functions, prerogatives, and accountabilities are usually codified in law, policy, or "agreements and understandings at [the] ministerial level."[2] Specific mandates may be assigned to a lead agency, with coordination requirements managed through other executive coordination processes or bodies.[3] In large, well-established democracies, multiple lower-level coordination bodies may manage sector- or issue-specific coordination. In such polities, there are normally clearly defined accountabilities for coordination bodies at all levels. Regardless of the size or number of actors, inter- and intraministerial coordination mechanisms are designed to overcome the compartmentalization of information and ensure effective, timely, responsive, and strategic decision-making.

Such coordination is particularly critical during crises, when decision-making timelines are compressed, information is imperfect, resources may be scarce or dispersed, and additional actors or equities outside of normal decision-making processes must be consulted or included. Although crisis management arrangements are usually not standing entities, they are nonetheless highly institutionalized to ensure rapid activation when required.[4] Activation may be triggered by events such as terrorist attacks, natural disasters, or mass riots for which the defense sector has a primary or supporting function. In weak or transitioning states, the defense sector may play an outsized role in internal security crises when internal security forces lack capacity or resources, necessitating certain emergency measures and a high degree of oversight and accountability.

Whereas interministerial coordination requires optimizing the coordination between the defense ministry and external agencies and government bodies, intraministerial coordination focuses inwardly to synchronize and coordinate functions essential for delivering the overarching defense mission. How strategic-level decisions, policies, and actions at headquarters direct, guide, or inform operations and how, in turn, operations feed into policy-level decisions are other essential components of effective defense sector management and coordination. Civilian policymakers that are isolated from the needs and conditions of their operational components may make decisions on materiel, weapons procurement, or training, for example, that do not serve the operational needs or strategic mission of the organization. In large, complex, or geographically dispersed defense sectors, defense sector coordination and management ensure that different components, which may be agencies or departments in themselves, are either centrally managed or otherwise answerable to the overarching mission of the organization. In less complex or already centralized defense sectors, this coordination ensures that different offices or divisions are not siloed and that their work contributes to the broader defense mission.

For both normal and crisis coordination to be effective, processes must be clearly defined and codified. Permanent arrangements and crisis response options must include identification of a lead body or agency, including assigned staff, and "procedures for coordination, information sharing, [and] resource allocation."[5] For each participating ministry or defense sector agency, there must also be rules and procedures outlining responsibilities, operating procedures, and information sharing requirements. These ensure a clear division of labor within the defense sector and between the defense sector and other ministries that is essential for effective government.[6]

Guiding Principles for the Design and Implementation of Goal 4

The following two principles provide guidance for how to achieve the goal of defense sector coordination and management across a wide range of contexts.

First, there must be a clearly defined, codified, and resourced mechanism for coordinating, at a minimum, defense and security matters that impact the mandates of other government ministries. Although the mechanisms will vary widely by polity, there should be a designated body, such as a national security council, or a lead agency that bears responsibility for coordinating decision-making, information sharing, resource allocation, and government response. In small or poorly resourced states, this designated body may be an individual, likely a minister or deputy minister, with a supporting staff. Regardless of the structure, this role, and the required mechanisms for interministerial coordination, must be clearly defined. Usually, this role is codified in policy or statute, although it may also function according to well-established precedent. Furthermore, these guidelines must address coordination requirements for both routine matters and crises. Finally, issues that always demand an interministerial response—such as a terrorist attack or natural disaster—either must be assigned to a department or agency lead for coordination in accordance with the standard guidelines or must be subject to issue-specific, location-specific, or other specific guidelines. To be effective, these guidelines must also detail responsibility and accountability for coordination and information sharing.

In weak or transitioning states, interministerial coordination mechanisms are often absent, weak, poorly resourced, or ad hoc. Transitioning authoritarian states frequently feature structures that centralize power and decision-making at the very highest levels of government. These are often accompanied by either written or unwritten prohibitions (or sanctions) for horizontal coordination (i.e., coordination with entities or individuals outside of the direct chain of command) absent direction from the central authority. Where such structures are long-standing, there is often also an institutional culture that eschews information sharing and the types of formal and informal structures that facilitate it—such as cross-ministerial committees or working groups, let alone formal decision-making councils. In some cases, defense ministry officials are required to report any interaction with officials from other ministries to internal security officers, with the effect that few such interactions occur. In these states, the mechanisms and institutional cultures designed to prohibit coup d'états present the greatest hurdle to establishing effective coordination following transition to democratic rule. Here, the challenge will not just be to establish those mechanisms but also to ensure that organizational cultures can evolve to support their effective use.

In small or poorly resourced states, the challenge may be one of numbers. There may not be enough personnel to take on the additional task of managing coordination between the defense ministry and other ministries. Or there may not be sufficient, or readily available, information with which to conduct coordinated decision-making, particularly in times of crises. Often, platforms are lacking to manage the flow of information or timely decision-making, particularly for dispersed agencies. Here, the challenge is not that the processes for interministerial coordination are missing or incomplete but that there is a lack of people or information to make even the most basic level of coordination possible. In such circumstances, it may be necessary to achieve the goal of interministerial coordination through an ad hoc mechanism (e.g., a coordination body activated by a senior member of government with named officials based on coordination needs) as a first step to creating a more permanent mechanism in the future.

Too often, defense reform efforts focus on the *mechanism* for coordination, such as establishing a new ministerial council or creating a counterterrorism coordination cell. What is often overlooked, or delayed until after the new body is created and staffed, are the specific rules and operational guidelines for how it functions. These guidelines can be tremendously difficult to create because decision-making authority for these issues already resides with a ministry, agency, or individual. Establishing new guidelines thus also means divesting key actors of those functions or otherwise streamlining decision-making in ways that reduce their power and authority. In many cases, these actors will either derail the creation of new coordination guidelines or otherwise act to ensure the coordination mechanism itself fails. Resources are withheld, staff are reassigned or "short cycled" (i.e., assigned for a brief period and then reassigned elsewhere), or requirements are effectively ignored or deliberately derailed. In other cases, the division of labor is not clearly defined, leaving key interministerial initiatives virtually deadlocked.

Second, clear policies, procedures, guidelines, and systems should be established to synchronize and coordinate activities, functions, and responsibilities among different offices, departments, and agencies within the defense sector. Although the type and number of these guiding documents, coordinating bodies, or established practices will vary greatly with the size, complexity, and dispersion of defense sector components, the central issue is that all defense sector components, no matter how large or far-flung, operate in accordance with and under the direction of defense sector leadership and in fulfillment of the defense sector mission.

Achieving the appropriate level of intraministerial coordination in large, well-established democratic polities may involve creating new coordination bodies; establishing mechanisms, guidelines, and authorities for information sharing and decision-making among defense sector offices or agencies; or developing specialized training programs for personnel to improve coordination practices. For example, personnel may be required to complete management and leadership training to join the executive ranks of the defense sector,[7] and these programs can be modified to include best practices for intraministerial coordination and management. Such efforts can involve defining the skills and training that make managers successful and identifying the elements of training programs that enable managers to effectively operationalize strategy.

In small or poorly resourced states, information may not be sufficient, or readily available, to enable coordinated decision-making to take place, particularly in times of crises. Often, platforms are lacking to manage the flow of information or timely decision-making, particularly for dispersed ministries. In many states, there is no official or secure email system, phone lines may be unreliable, or electricity may be lacking to power computers. Here, the challenge may be to develop the infrastructure to make even the most basic level of coordination possible. Similarly, standard processes such as budget cycles and procurement practices may be nonexistent. As a result, budgets are allocated at the top without much consideration for the needs and requirements of the defense ministry as a whole. In other instances, procurement of goods happens in one office, but the office that will receive the goods is unaware that a shipment is coming. These examples are illustrative of the organizational silos, duplicative efforts, and lack of coordination that hamper the effectiveness of the defense sector and thus its ability to fulfill its mission. Where personnel changes occur frequently, simple steps, such as disseminating an organizational chart with the names and contact information of key decision-makers within the organization, may be a first step toward ensuring that ministry personnel know who works across the sector and whom to contact to coordinate and execute decisions.

Transitioning authoritarian states often have the required infrastructure but lack the institutional culture and trained personnel to reform processes and structures that were likely designed for the compartmentalization, rather than the sharing, of information and decision-making. These challenges may require enhancing transparency across defense sector offices and agencies, providing training for managers and leaders on how to coordinate decisions and build consensus, and establishing other mechanisms to introduce a gradual shift in the organizational culture of the defense sector. The challenge in such polities is often one of human rather than institutional capacity. Mechanisms and guidelines exist, but personnel are either unaware of these processes or afraid to use them.

According to the PAP-DIB Sourcebook, establishing intraministerial coordination mechanisms requires that there be "a clear division of responsibility in missions . . . and clearly defined obligations of accountability."[8] The absence of such mechanisms is reflected in overlapping missions, budget and resource redundancies, and ambiguous attributions of responsibility and accountability. In other words, there should be a clear delineation of where the authority and the responsibility lie for policy decisions, strategy-making, procurement actions, and operational orders at the ministry level and the mechanisms by which operational components have input into, or are otherwise informed in a timely manner of, such ministerial-level actions. At the same time, it is equally important that there are clear guidelines for when and how, and at what level, operational decisions and actions must be coordinated with ministerial decision-makers. Both components—the policy and the operational—have a role to play in the execution of the defense sector's mission, and neither can operate effectively in a silo or in conditions where the division of responsibility for timely information sharing and coordination of decisions is not clearly delineated or otherwise codified. There should be no question, either before a decision is to be made or after the fact, of which position or entity within the defense sector has the authority and responsibility for sharing information or coordinating decisions and actions.

Not surprisingly, addressing this challenge may be more complicated in polities where multiple actors have overlapping interests or roles in the matter at hand and where the defense sector is large, complex, or dispersed. In such cases, defining who is responsible for what and when is likely the best place to start unraveling overly complex systems with too many decision-makers and a lack of clear authority. On the other end of the spectrum, highly siloed or centralized defense sectors that are tightly managed or controlled by just a few decision-makers face an entirely different set of challenges. Here, the task will be to loosen the tight hold on decision-making, devolve authorities to defense sector commands, enhance transparency and information sharing, and promote wider input into key decisions and actions between ministry leaders and operational commands. Although the complexity of the challenge is less, the solution still involves clearly defined roles and responsibilities and assigned authorities for coordinating decision-making and actions between ministry and operational actors.

Improving intraministerial management and coordination, regardless of size and type of polity, will involve streamlining or reorganizing cumbersome or bloated bureaucracies. These heavily siloed institutions often struggle to manage the timely sharing of information, coordinate decisions, or execute actions within the defense sector. Replacing outdated structures designed to prevent information sharing, as well as dispensing with written and unwritten sanctions for working across offices and functions, may also be required. Attention must also be paid to how such organizations will incentivize personnel to support new mechanisms and abide by new guidelines.

Effective coordination and management require that both the mechanisms and the guidelines be established. One without the other may leave the defense sector either less capable or, in other cases, exclusively in control of coordinating matters vital to the country's defense, security, and foreign policies and interests.

Applying the Goal 4 Principles in Practice

These two principles provide conceptual guidance for how to achieve the goal of defense sector coordination and management in a wide range of defense sector reform contexts. Their application in practice—particularly when designing defense sector reform activities or interventions—is a contextually driven exercise. Under each guiding principle, some examples of how these principles might generate activities or interventions are provided to help guide the practitioner to translate them into possible activities in a specific defense sector reform environment.

1. A clearly defined, codified, and resourced mechanism defines how defense and security matters are coordinated with other government ministries.

The first step in designing defense sector reform interventions to address interministerial coordination is to map what processes exist, if and how those are codified, and whether and to what extent there is adherence to that process in practice. Although the mechanisms will vary widely by polity, a body, such as a national security council, or a lead agency that bears responsibility for coordinating decision-making, information sharing, resource allocation, and government response, should be designated to coordinate between the defense ministry and other government ministries, and the required mechanisms for interministerial coordination must be clearly defined and ideally codified in policy or statute.

The example of Georgia is illustrative. In March 2015, the Georgian parliament passed the Law of Georgia on Planning and Coordination of the National Security Council, which established the policies and procedures for the planning and coordination of national security policy, including defense, external security, international security, social and economic security, ecological and energy security, and information security.[9] The law designated a consultative body, the State Security and Crisis Management Council, which is accountable to the prime minister, to exercise the principal supervisory role for coordinating the preparation of the draft national-level conceptual documents in the areas of state defense, security, and legal order, and for monitoring the implementation of the action plan of the National Security Strategy.[10] The council included broad representation. Chaired by the prime minster, the council convened six times a year. Permanent members included the minister of finance, the minister of internal affairs, the minister of defense, the minister of foreign affairs, the head of the State Security Service, the assistant to the prime minster for state security issues, and the secretary of state. Ministry-specific regulations also exist for each of the agencies represented by permanent members on the council, defining their roles and responsibilities and regulating interagency activities for managing defense. Nonetheless, vague language and the overlap of responsibilities between the National Security Council and the State Security and Crisis Management Council created gray areas in the legislation that enabled government entities, such as the Ministry of Defense, to bypass coordination requirements.[11] In 2017, new legislation removed the duplicative roles of both bodies, but without designating a replacement body, leaving Georgia with a "gap in the fields of planning and coordination of the national security policy as well as the adoption and coordination of strategic-level decisions in all types of crisis situations."[12]

Where codified mechanisms for coordination exist, as in the example of Georgia, this mapping will require close attention to vague statutes and gray areas in roles and responsibilities that undermine the effectiveness of the coordination mechanism. Likely defense sector reform interventions could focus on addressing those areas by advising revisions to legal statutes or by providing procedural improvements or training for relevant staff on adherence to existing statutes. Where challenges are generated by poor information sharing, interventions could be geared to more basic technical assistance—for example, designing mechanisms for better or more timely information sharing between the lead coordinating agencies and other relevant government agencies and ministries. In Mali, for instance, interministerial coordination is highly fragmented. The system for sharing timely information, particularly in times of crisis, and for coordinating a response does not function effectively among government institutions or even across geographically dispersed entities within those institutions. Further challenges result from the poor quality of information that is shared, further impacting the coordination of decision-making.[13]

In other instances, interministerial coordination may be stymied or blocked by legacies of institutional distrust or by ministerial resistance to coordinating mechanisms that reduce their power to act unilaterally or their influence over governmentwide decisions. In Tunisia, for example, long-standing tensions between the Ministry of Interior and the Ministry of National Defense have derailed efforts to generate greater interministerial coordination—as seen when a donor-funded initiative to enhance counterterrorism coordination was deadlocked until the coordinating entity was established under the control of the Ministry of Interior. Transitioning authoritarian states frequently feature structures that centralize power and decision-making at the very highest levels of government. These are often accompanied by written or unwritten prohibitions for coordination absent explicit permission. Where such systems are long-standing, there is often also an institutional culture that eschews information sharing and the types of formal and informal structures that facilitate it. In such contexts, defense sector reform interventions may be impossible without addressing those institutional cultures, because without such efforts, new coordinating entities can be generated on paper but will have little practical impact without the effective participation of the entities they are supposed to coordinate.

Where programming does seek to generate institutional cultural changes—for example, through training, advising, or embedded assignments of key personnel—those efforts may take years to have any impact. Often, real change is not accomplished until a new generation of leaders takes the helm and commits to genuine coordination. In the meantime, efforts can generate technical or tactical coordination on critical issues, particularly when addressing those issues is in the interests of the coordinating ministries or their lead staffs. For example, in Tunisia, the rapidly escalating terrorist threat in 2013–14 generated the push for interagency coordination on efforts related to counterterrorism as an entry point for addressing other governmentwide coordination issues, because the issue was widely seen as an existential threat.

Alongside generating shifts in institutional culture, other near-term initiatives can focus on removing authoritarian-era sanctions on interministerial discussions among senior staff or revising information classification systems to enable information sharing between government entities on matters of security and defense, which in some postauthoritarian contexts may be largely or entirely prohibited.

In Iraq, the challenge is compounded by competing sectarian loyalties. The Iraqi National Security Council is chaired by the prime minster. Other members include the national security advisor, the minister of defense, the minister of the interior, the minister of finance, the minister of foreign affairs and the minister of justice. In addition to their political affiliations, each official also advocates for their respective sectarian interests. The existence of "competing armies" further compounds interministerial coordination.[14] The Iraqi Army's mandate competes with the Counter Terrorism Service, the Federal Police (including its Emergency Response Division), the Border Guard Force, the

Energy Police, the Popular Mobilization Forces, and various tribal forces, and similar competition exists for scarce resources. Some estimates place the number of Iraqis serving in these forces as high as 750,000, effectively crowding out funding for sustainment and modernization.[15] The Office of the National Security Advisor is designated to serve as the coordinating body between the different components of the ISF but is dominated in practice by sectarian rivalries. According to the Iraqi National Security Advisory's Coalition Provisional Authority Order No. 68, "the Ministerial Committee for National Security is the main forum for taking decisions related to these issues at the ministerial level."[16] In practice, however, each force competes for its own interests and for the largest share of defense spending. As a result, the United Nations Assistance Mission for Iraq has maintained a position for interministerial and intergovernmental cooperation and coordination to address the absence of coordination and management among government bodies with a security mandate.[17] In postconflict contexts, similar positions under international mandates may be required to serve as coordination mechanisms until these can be created or made effective as the country transitions from conflict.

2. Clear policies, procedures, guidelines, and systems coordinate activities, functions, and responsibilities among different offices, departments, and agencies within the defense sector.

Achieving the appropriate level of intraministerial coordination in large, well-established democratic polities may involve creating new coordination bodies; establishing mechanisms, guidelines, and authorities for information sharing and decision-making among defense sector offices or agencies; or developing specialized training programs for personnel to improve coordination practices. For example, a 1998 law in Colombia defined the managerial functions of the minister of defense, which were subsequently restructured under Decree 1512 of 2000.[18] A general management unit within the Ministry of National Defense has responsibility for developing ministerial manuals, including the development of a system for managing quality performance that each defense sector entity must implement. This guidance complements the ministry's code of ethics, which focuses on raising awareness among public servants about their duties and responsibilities and establishes the essential values that should guide the institution.[19] Although the type and number of guiding documents, coordinating bodies, or established practices will vary greatly with the size, complexity, and dispersion of defense sector components, the central issue is that all components of the defense sector, no matter how large or far-flung, operate in accordance with and under the direction of defense sector leadership and in fulfillment of the defense sector mission.

In small, poorly resourced, or postconflict states, information may not be sufficient or readily available to make coordinated decision-making possible, particularly in times of crises. Often, the challenge is compounded by the existence of various armed factions, militias, and other tribal forces, some of which draw on the ministerial payroll but are not under the defense sector's effective command. In Iraq, one of the principal challenges for coordination within the defense sector is the existence—as noted above—of "competing armies." Libya faces similar challenges. Following a successful conclusion of peace accords in Libya, intraministerial coordination and management mechanisms will need to be rebuilt to integrate forces serving under Libya's Tripoli-based GNA and the Haftar-led Libyan National Army, while also potentially incorporating former armed group fighters under a possible disarmament, demobilization, and re-integration (DDR) program.[20] This will be no small task. Decision-making in the defense sector is severely impeded by a lack of centralized, ministerial-level authority; ineffective communication; the absence of clear chains of command; and the influence of affiliated armed groups, over which the defense sector has little if any effective control.[21] Most armed groups under the Ministry of Defense continue to maintain autonomy, and the ministry is often bypassed by direct reporting relationships between the GNA's executive branch and armed group commanders.[22]

The overlapping mandates of Georgia's security forces present yet another defense sector reform challenge. Georgia's primary external security and defense force, the Georgian Armed Forces, is under the authority of the Ministry of Defense. However, other forces with overlapping mandates include border guards, interior troops, and the State Security Service, with the effect that "some military and police functions are mixed without any clear division of responsibility."[23] The general staff is entrusted with a coordinating role over all military forces in times of mobilization or martial law. Nonetheless, Georgia's Law on Defense does not clearly distinguish the functions and levels of coordination between defense and internal security forces. Additionally, interior troops have been declared part of Georgia's military forces while remaining subordinate to the minister of interior, "violat[ing] the constitution, which categorically forbids the merger of defense, police and security forces."[24]

Designing defense sector interventions to address the absence of intraministerial coordination and management in such varied contexts will be tied in large part to postconflict peace processes and possible DDR programs, particularly where nonstatutory armed groups fulfill both sanctioned and unsanctioned defense sector operational mandates, or to ongoing reform of legal statutes, including changes to the constitution, where mandates are not clearly defined. As part of legal reform initiatives, such as a postconflict constitution drafting process, efforts can focus on creating new coordination bodies or reforming pre-conflict mechanisms that still exist or have some legitimacy. Other efforts can focus more narrowly on technical mechanisms for coordination, particularly where larger political processes are still unfolding to address near-term management needs. Still other near-term efforts might focus on training to build the defense sector's human capital for coordination and management in anticipation that new processes being established will need personnel to staff them.

Where political buy-in for coordination and management reform is lacking, or where peace processes are still in their infancy, other options that are narrow in scope or highly technical in delivery might also be possible, depending on the context. For example, in Iraq, payroll processes still require hand delivery of cash to deployed commands as the only method to pay deployed soldiers.[25] Establishing alternative payment methods that are less susceptible to corruption or that ensure forces are being paid might be a critical early step toward retaining the loyalty of forces during the postconflict transition. Even in these initial stages, attention must be paid to how personnel will be incentivized to support new mechanisms and abide by new guidelines.

CHAPTER SEVEN

GOAL 5
Functioning Logistics

The fifth goal of defense sector reform is functioning logistics. Logistics can be defined as "the aspect of military science dealing with the procurement, maintenance, and transportation of military materiel, facilities, and personnel" or, more simply, as the "practical art of moving armies and keeping them supplied."[1] To be "functional," a country's defense logistics must be able to get the right things and people to the right place at the right time. Logistics processes, resources, and systems are the cornerstone of an effective defense sector, and their collective functionality has a direct impact on the warfighting capability, agility, and range of movement of the armed forces.

At a basic level, logistics deals with the core functions of generation, deployment (and redeployment), supply, sustainment, and transportation. *Generation* is the process by which the right resources (human, materiel, and services) are designed, developed, and obtained to provide commanders with the necessary capabilities to accomplish their mission.[2] Generation is intimately tied to a force's readiness. *Deployment* refers to the function of preparing and arranging for the deployment of personnel and materiel to an area of operations; *redeployment* refers to the functions involved in evacuating materiel and military personnel for the purposes of maintenance, reconstitution, and, importantly, medical care.[3] *Supply* refers to the function of providing, warehousing, storing, and maintaining the materiel and personnel needs (food, water, fuel, ammunition, and other goods and services) of the military forces. *Sustainment* is the function of maintaining what the force needs to operate in theater until its mission is achieved. *Transportation* is the movement of materiel and defense personnel into, throughout, and out of a theater of operations.[4]

Many advanced democracies have additional activities on their list of core functions based on the complexity of their operations; the United States, for example, includes engineering and operational contract support among its core logistical functions. In such large and well-resourced states, logistics are delivered by highly complex logistical systems that span the globe, are sometimes multinational, and often rely on sophisticated information management and information technology, whereas in much smaller and more poorly resourced states, logistics may be rudimentary, relying on analogue systems, such as inventorying using index cards.[5] Regardless of how advanced the country's defense capabilities are, a functioning logistics system should at least be able to support and provision its military to execute the missions identified by the defense sector's strategy and plans. "Functionality" does not require a specific type of system or a certain level of complexity; it requires only that the system is able to identify, plan for, and process needs; move required personnel, equipment, and other items to meet those needs; and respond to ministerial directives or field-based requests for resources in an efficient and timely manner.

Among countries where defense sector reform is a priority, what constitutes "logistics" may vary significantly. In some countries, the logistical system will be a product of various legacy systems and contributions from foreign governments. A country may have, for instance, a bureaucratic system for logistics management dating back to colonial rule, a Soviet-era system for tracking and repairing equipment, and a NATO system for coding parts and equipment. Variations will also depend on the geographic reach and complexity of the forces the logistical system is designed to support; countries with multiple armed services that operate globally are likely to have far more complex logistics than a small landlocked country that only has, for instance, an army and a small air force. What comprises core logistical functions also varies by country. For example, the provision of medical supplies, the movement of medical support, and patient or casualty evacuation fall under the purview of logistics in some countries but not in others. The repair, support, and modernization of equipment is another function that is part of logistics but also part of procurement, particularly when the equipment, as well as spare parts and the training to repair that equipment, is procured from foreign countries. Despite some important variations, most logistics systems include the management of materiel (e.g., introduction of new materiel, use, control, and disposal); management of facilities (e.g., use and maintenance); management of transportation (e.g., national and international distribution and movements); and management of services (e.g., contract arrangements, setting food and other standards).[6]

Implementing defense sector reform to achieve the goal of functioning logistics requires identifying why the existing logistical systems are not functioning. Dysfunction may derive from an inability to track equipment and spare parts because there is no basic inventorying system, no warehousing system to store needed parts, or no tracking system to know where materiel is at any given point in the supply chain. In other cases, the breakdown may result from the lack of a system for planning for longer-term needs and responding to urgent requests. Systems for communicating logistical needs from the field may also be missing. In some cases, logistics systems may be dysfunctional by design. Inefficient logistics systems provide ample opportunities for corruption, and fixing those systems may generate significant resistance from those who stand to lose from a more formalized and process-driven system.[7] Because functioning logistics requires systems, a breakdown in any point in the logistics system can lead to systemwide dysfunctions, with severe implications for military effectiveness and national security objectives.

Despite the centrality of logistics for the effective deployment and sustainment of military forces, relatively little attention is given in the literature to how defense sector reform programmers can actually develop logistical capacity. The UN Defense Sector Reform Policy lists among its main areas for support the development of "sufficient national governance, management, institutional, resource (human, materiel and financial) and technical capacities and capabilities, in the strategic, operational and tactical dimensions of a national [defense] sector," but fails to provide specific guidance for how the capacity associated with functioning logistics can be implemented.[8] Similarly, the U.S. Department of Defense's DIB Directive specifically includes logistics as one of the "principal functions and duties of an effective defense institution" but provides little detailed programming advice for implementers who seek to improve host-nation logistical capacity at the institutional level or otherwise.[9]

The NATO PAP-DIB Sourcebook addresses logistics more directly. When discussing how security sectors should function, logistics is recognized in the sourcebook as an essential function for military preparedness and intrastate cooperation.[10] It also acknowledges the importance of logistics to supporting tasks, coordinating across other training and engagement activities, performing calculations, and supporting systemwide processes.[11] In noting how some governments have privatized logistics and maintenance while others have nationalized production systems, the sourcebook also discusses how logistics is tied to government transparency and larger topics of financial planning and defense spending.[12] However, while logistics is largely discussed as an assumed capability or emphasized as important, the sourcebook also largely falls short of describing what processes are required to institutionalize and integrate systems for each of the five basic core functions for defense logistics to operate effectively.

Guiding Principles for the Design and Implementation of Goal 5

Functioning logistics is critical to the efficacy of any defense sector, but in countries where defense sector reform is likely to be a priority, the type of systems, their degree of sophistication, and their general level of interoperability will vary significantly. How, then, can the goal of functioning logistics be realized? The following two principles can help inform how to achieve that goal across a wide range of contexts.

First, functioning logistics requires effective systems for each of the core components of logistics: generation, deployment, supply, sustainment, and transportation. While each country's logistics will vary, there must be systems in place that ensure the defense sector's broader logistical framework is able to (1) design, develop, and acquire or construct the right materiel and services resources; (2) provide, warehouse, store, and maintain those resources; (3) move human and materiel resources to where they need to be when they need to be there; and (4) prepare and arrange for the deployment of resources to an area of operations, and then out again for the purposes of maintenance, reconstitution, and medical care. Although logistics can include many additional functions, functionality requires at least these.

Generation refers to the function that includes "the production and procurement of military forces and serves as the foundation of military logistics."[13] In its most basic sense, generation is about readiness; it includes all the processes by which the military ensures that it has the right people and equipment ready to address current and future operational requirements. In the U.S. Army, for example, generation includes processes that "achieve progressive levels of readiness with recurring periods of availability as both active and reserve component units progress through three distinct force pools: RESET; Train/Ready; and Available."[14] The first of these generation "pools," RESET, includes the troops and equipment that have recently returned from deployment. During this phase, units restore their people and resources to a sufficient state of readiness that the units can begin training again. In the second phase, "Train/Ready," troops "receive new personnel, manage and retool equipment, and begin collective training." And in the final phase, "Available," troops and materiel resources either deploy for rotation on current missions or remain actively available for contingency missions.

Deployment refers to the functions of preparing and arranging for the deployment of personnel and materiel to an area of operations. *Redeployment* refers to the functions involved in evacuating materiel and military personnel for the purposes of maintenance, reconstitution, and, importantly, medical care. In military logistics, deployment systems determine how materiel and forces will be distributed and facilitate that distribution.

In defense logistics, *supply* refers to the materiel items used to equip, support, and sustain defense forces. Supply includes the design and development of military materiel and services, as well as their manufacture, purchase, and procurement. Supply also includes the storage, distribution, maintenance, repair, salvage, and disposal of materiel. The supply function serves to support the military forces to "live (food, water, clothing, shelter, medical supplies), to move (vehicles and transport animals, fuel and forage), to communicate (the whole range of communications equipment), and to fight (weapons, defensive armament and materials, and the expendables of missile power and firepower)."[15] Supply systems vary depending on whether the resource is singularly expendable/consumable (such as food, fuel, and ammunition) or reusable (such as uniforms and weapons). In advanced military systems—such as the U.S. armed forces or NATO—supply systems are highly complex with multiple subsystems. The United States, for example, recognizes ten individual supply classes (e.g., ammunition, vehicles, medical materiel, personal demand items, and repair parts), each of which have their own supply and sustainment systems that integrate into the larger logistics system, while NATO has five classes that include the same supplies in a different configuration.[16]

Sustainment is intimately linked to supply and refers to the functions necessary to support the armed forces to maintain their combat power until their mission is accomplished. It is not enough for a defense establishment to supply its forces; it must also have systems in place to ensure sustainment of the forces during operations. Sustainment relies on what the U.S. Army terms the "Principles of Sustainment," which include logisticians' ability to anticipate need by forecasting what the forces will need to achieve their missions; respond to changes on the ground; simplify processes and procedures to minimize complexity; economize resource provision to ensure efficient management and usage; maintain survivability, which involves ensuring sustainment despite hostility in the environment; maintain continuity of providing sustainment across all levels of war; integrate all sustainment elements within operations; and improvise to adapt to unexpected conditions during a mission.[17]

In defense logistics, *transportation* systems (sometimes referred to as "transportation and movement") include all of the airlift, ground transport, sealift, infrastructure, facilities, command and control, and equipment involved in the flexible deployment, movement, sustainment, and redeployment of the defense forces to, within, between, and out of theaters of operation.[18] More simply, a functioning logistics system must have a system to transport people and materiel in a timely manner to ensure that the force can execute its mission. This capability involves both the straightforward movement of people and materiel from one place to another and more complex movements such as transporting materiel from foreign suppliers or forward-basing units and materiel to locations that span the globe.

In many of the countries where defense sector reform will be a priority, the five functions above are rarely systematized and those that are, are often dysfunctional. For example, while the generation function is well understood and integrated in advanced democracies, in countries in need of defense sector reform, the force structure and available resources will not reflect what the military needs to carry out its current missions (including adequately defending the country's borders) or to maintain a state of readiness for future or contingency missions. Generation requires forward planning. Countries that are mired in ongoing conflict often cannot spare the resources such planning requires. Generation also requires resources and personnel who are not deployed to be preparing for deployment. Defense sectors often lack the kind of system that will keep troops at a specific level of training. In Mali and Libya, for example, security forces have been expected to "learn on the job." In Libya, the closure of military training facilities has made learning on the job the only way to prepare forces.

In the case of supply, the problem may be one of outdated, inefficient, or nonexistent inventory management. The inability to effectively track defense resources is often directly related to other problems in the defense sector, such as ineffective resource allocation, strategy, and planning. In some countries, the defense sector may not know where its resources are, or their quality or quantity, and processes and procedures for tracking resources do not exist. Even in advanced democracies, inventory management is an area prone to fraud, waste, abuse, and mismanagement due to overstated requirements, poor oversight, and weak financial accountability. The acquisition of resources that the military does not actually need is also problematic and is often tied to fraud or corruption. Moreover, not knowing what supplies are available makes it almost impossible for commanders to plan effectively and puts the lives of deployed forces at risk.

Second, functioning logistics requires that the defense sector's functional logistical systems be integrated and coordinated. Although a country may have systems for each of the logistical functions discussed above, their existence alone is not enough to create functionality. To achieve that, the defense sector must have mechanisms in place to ensure that all functions within the broader defense logistical system are integrated and coordinated and work in unity to meet the logistical needs of the force in an efficient manner. Working in unity means that these systems must be integrated from the ministerial and general staff levels down to the military services and deployed units, such that each level supports and synchronizes with the other levels.[19]

Although the complexity and sophistication of an integrated logistical system can vary widely—from the most technologically advanced computer-based systems to simple analogue, paper-based systems—it is important that these coordinated systems ensure that (1) top-level planning is translated into the appropriate logistical support for operationally deployed units and personnel; and (2) the logistical needs of those units and personnel are communicated from the bottom up through the chain of command. The systems and processes used by the defense sector's logistics system must be interoperable and coordinated across the defense sector—both horizontally across defense sector institutions and services and vertically between field and headquarters. Units at the tactical level must be able to request replacement parts to fix and maintain their weapons systems. Simultaneously, commanders at the operational and strategic level need to be able to monitor the logistical status of the tactical units and to understand likely logistical challenges in order to be able to effectively adapt defense plans as conditions on the ground change.

The importance of integration and coordination cannot be overstated. To get a single piece of equipment to a deployed unit, for example, requires planning for the type of equipment required based on anticipated need; the existence of communication mechanisms for the commander to request the equipment; the development and production, acquisition, or inventorying (if it is being stored) of the piece of equipment; the transportation of the piece of equipment to the site of operations; the maintenance of the piece of equipment during its deployment; the retrieval of the equipment when it is no longer being used or needs to be repaired in a depot; and so on. Although each step in the process involves a different functional system, all the functions must be coordinated for the desired end result (getting the right thing to the right place at the right time) to be achieved. Every step in the integrated logistics system is essential not only for the execution of a force's mission but also for ensuring effective planning, the efficient use of materiel, and the prevention of corruption.

In countries where defense sector reform will be a priority, this integration of logistical functions is likely weak or nonexistent. In some countries, there may be a mix of formal systems for some functions, but weak, absent, or informal systems for others, further complicating their integration. Often, the integration of logistical systems will be problematic because of the different kinds of equipment and systems currently in use—often the result of military occupation by or assistance from a variety of different countries—none of which use similar parts or interoperable processes and procedures. In former authoritarian countries, logistics systems may have been deliberately siloed as a coup-proofing mechanism, whereas in postconflict environments, the lack of integration will stem from a broader breakdown of coordination across the entire defense sector. Because of the variation and complexity across contexts, the programming involved in improving integration will range from relatively straightforward assistance—such as the development of a streamlined process for a specific logistical support function or inventory tracking system—to more complex tasks, such as integrating existing systems with a new one, modernizing systems that are outdated or inefficient, or enhancing access to information for all users. Undergirding these challenges may be urgent requirements to improve lines of communication, particularly between headquarters and frontline forces.

Applying the Goal 5 Principles in Practice

Having functioning logistics is essential for the effective organization, training, equipping, deployment, and employment of the armed forces. The two principles provide conceptual guidance for how to achieve this goal in a wide range of defense sector reform contexts. Their application in practice, however, will be a contextually driven exercise. Under each guiding principle, some examples of how these principles may generate programming are provided to help guide the practitioner in translating them into possible activities across varying defense sector reform environments.

1. Established systems deliver the five core logistics functions of generation, deployment, supply, sustainment, and transportation.

Sourcing, maintaining, and transporting human and materiel resources are critical to ensuring that the armed forces can execute their missions. To perform those tasks effectively requires systems that rely heavily on formal processes and procedures, such as those that track what resources the defense sector has, where those resources are located, how to acquire new resources, what transportation is available for moving the supplies, and so on. Defense sector reform programmers are likely to find that such systems lack certain functions, are inefficient and informal, or simply do not exist.

In some cases, core logistical functions will exist but will be incapable of effectively providing the logistical support that the forces need to achieve their defense missions. In the case of sustainment and transportation functions, for instance, the problem is often one of resource availability. There may not be enough equipment, personnel, or transport to move what is required to the right location and at the right time. During the 2012–13 conflict in northern Mali, the Malian Armed Forces had to conduct missions in areas of the country occupied by Tuareg and Islamist insurgents, where the military had little to no permanent infrastructure. Expeditionary logistics were necessary to ensure the deployed forces were supported and sustained. Supply routes were established to move supplies (water, food, ammunition, and fuel) from the capital to the

north over some 750 miles of harsh, sandy terrain in searing heat.[20] This constrained the number of troops that the Malian Armed Forces could send north
to the level of logistical support that the logistical units could provide. When
the defense sector's supply system failed to provide the necessary equipment
and ammunition and its transportation and sustainment systems were unable
to move resources and troops to the north in a timely manner to sustain them,
soldiers mounted a coup against the sitting government, citing incompetence.
As one official has commented: "The failure of the Malian Army's operations
lay not primarily with the will or courage of its soldiers, but in deep systemic
flaws in the institutions of the military as a whole. Soldiers fighting in the North
quickly ran out of bullets and food because of the lack of a logistics system
capable of resupplying them."[21] Without essential equipment, sufficient ammunition, or an adequate number of troops, the Malian Armed Forces suffered
significant setbacks in 2012 at the hands of the insurgent troops, including the
loss of control of the country's three major northern cities.

In other cases, countries depend on foreign or contracted logistics in place
of developing their own core logistical functions. This reliance is often the result, in part at least, of security assistance programs that deliver equipment but
fail to pay adequate attention to the functional logistics systems that are essential to their sustained use. Without them, equipment quickly becomes unusable
and the country becomes more reliant on outside or expensive contracted maintenance. In Iraq, for example, substantial equipment was provided to help the
ISF, but attention was not paid initially to the development of functional Iraqi
logistical systems. Instead, donors relied on contracted logistics to support the
forces, which ultimately had a significant and detrimental effect on the forces'
warfighting capability. The serious implications of relying on contracted logistics instead of developing core logistical functions was evident during the ISIS
push through Anbar Province.

> Several officers said the system the Interior Ministry had devised to
> supply its forces was suited for peacetime, and predictably failed in
> war. They said it relied on contracts with businesses that would de-
> liver supplies to the troops' main garrisons. But as the border-police
> convoys headed for territory under militant influence or control, the
> vendors would not follow. "When the conditions were not good,
> and there was no security, they did not provide water and food by the
> contract," said a border-police colonel, who asked that his name be
> withheld. By June 17, the brigade was in position around al-Qaim,
> with hopes of blocking ISIS fighters' free passage to and from Syr-
> ia. But supplies were so depleted the troops could barely fight. Its
> members said they were given only a small piece of cake and about
> 10 ounces of water a day. Morale sank further, brigade members

said, because another unit in al-Qaim had already run out of water
and food, prompting the Ninth Brigade to share its meager stores.[22]

Contracted logistics providers can choose whether they want to enter a combat
zone. In this case, the contractors refused to bring their desperately needed
supply to the frontline Iraqi units. Given the importance of logistics to the defense forces, outsourcing logistical functions in place of developing the capacity
within the defense sector is not only inefficient but also dangerous.

In cases such as Mali and Iraq, the initial challenge for implementers will be
to determine which logistics functions exist and which core systems are missing
and then to understand how those existing systems coordinate (if at all) with
one another. It will be important to understand why systems that exist are not
functioning effectively before an appropriate solution can be devised to address
any dysfunctionality. Although there is often "a great temptation to declare that
a host nation's logistics system is so poorly organized that the only solution is
to completely replace it with a new one," this approach often results in "making
an already poor logistics system worse."[23] The country's forces will be relying on
whatever functionality the existing logistical system can provide, and removing
it completely in favor of a wholesale overhaul could endanger troops. Instead,
programmers should try to understand exactly which points in the system are
causing a breakdown in functionality and why. In countries such as Mali, this
may reveal inefficiencies in transportation or inventory management systems
that could be improved by streamlining them and training their personnel in
how to operate these newly streamlined systems. In other cases, the challenge
may be transforming informal systems into formal ones by codifying procedures, roles, and responsibilities. In countries such as Iraq, the focus may be
on developing each of the core logistics functions from the ground up, while
allowing for a gradual transition away from external contracted support as the
military manages to build its own capacity to carry out the core logistical functions itself.

2. Core logistical systems are coordinated and integrated.

Functioning logistics requires that the defense sector's logistical systems are integrated to ensure that forces get what they need when they need it to successfully execute their missions. The degree of integration in countries targeted for
defense sector reform can vary widely, from postconflict environments, such as
Libya, which effectively has no functioning logistics system or capacity, to relatively advanced democracies, such as Colombia, which has a well-established
logistics system that is one of the 11 subservices, or "arms," of the Colombian
Army. At one end of this spectrum, recipient countries will require programs to

develop all the logistics functions and the mechanisms to integrate them from scratch, whereas at the other end, programs will be far more specific and may focus on one single problematic element that is impairing coordination within the system as a whole.

In some cases, a lack of integration in the logistical system may be tied to poorly blended logistical system types (e.g., logistics systems based on Russian models but blended with those developed in North America) and to the presence of equipment from multiple origins. Military equipment from a single origin is usually designed to interact and work with other equipment from the same source. When equipment is sourced from many countries across several time frames, it will have different parts and require different methods of repair or will require advanced maintenance systems to which the country does not have reliable access.[24]

In Iraq, for example, coalition forces began to focus on building a logistical system for the ISF that would shift the forces toward self-reliance for their logistical needs. However, the ISF's logistical system comprised (and still does) a dizzying array of equipment, largely provided to the government of Iraq through foreign assistance. This widely varying equipment makes it almost impossible for Iraq's logistical system to maintain a stock of repair parts.[25] As of January 2021, for example, the United States had "$19.7 billion in active government to government sales cases with Iraq under the Foreign Military Sales system" and had "provided Iraq numerous systems under the Excess Defense Articles program, which offers excess U.S.-origin military equipment to allies and partners on a grant or sale basis, in an 'as is, where is' condition."[26] However, Iraq continues to purchase military equipment from other countries, such as Russia, China, and NATO partners, to augment the ISF's existing equipment. This adds to the burden placed on the struggling Iraqi logistical system by forcing the supply chain to reach back to numerous countries, vendors, and partners. The result is an overly complex and highly inefficient system of logistics that greatly hinders the performance of all elements of the Iraqi defense sector.

The example of Iraq is by no means unique. Implementers will have to devise solutions to the complex challenges Iraq exemplifies in numerous defense sector reform environments. There is no easy way to create an integrated logistics system when the current system has so many noninteroperable parts. A starting point will be to make sense of what systems, equipment, and weapons the country has in its logistical "arsenal," from which more specific entry points can be identified. In the case of Iraq, for example, one entry point for building integration was improving communications mechanisms among the suppliers, the logisticians, and the end users. By developing these lines of communication, greater coordination was both possible and increasingly institutionalized, which is essential for integration across the various logistical functions. Other entry points may be an inventory or maintenance system for the complex array of equipment in use; countries often need to know what they have, or ensure what they have is usable, before transporting it to where it is needed or developing a capacity for longer-term planning. In many instances, initial entry points will be dictated by needs on the ground, and programmers will have to focus their initial assistance on interventions to address the most urgent gaps before developing systems for longer-term sustainment. Nonetheless, understanding why the system is not functioning will be essential to devising any solution beyond the immediate future.

In countries where violent conflict is ongoing or has ended only recently, an overarching, integrated, and coordinated logistical system may not exist. In the case of Iraq at the outset of the war in 2003, for example, the security forces did not have their own established logistical system, and the ISF relied on contracted civilian companies for logistics. Initially, the lack of efficient logistical support stemmed from a combination of major combat operations, an inefficient theater setup, insufficient transportation capacity, and a lack of adequate theater distribution planning tools.[27] Rather than see the ISF fail in combat operations due to a lack of logistics, U.S. commanders provided all classes of supplies to their Iraqi counterparts, but as the ISF grew in capacity, the United States was forced to assist in making up for the ISF's lack of logistical capacity by outsourcing logistics to contractors. The result was an ISF that depended on U.S. assets for nearly all their logistical needs; even Iraqi maintenance depots were colocated with U.S. logistical units (both contracted units and U.S. military units). As U.S. forces drew down from Iraq, the ISF's logistical systems were neither efficient nor effective, and the ISF modeled its approach on the American one and contracted logistics.[28] This overreliance on contracted logistics to supply frontline tactical troops, coupled with the lack of a functional logistical system, ultimately resulted in logistical and, in turn, mission failures for the ISF.

In countries such as Iraq that have established a reliance on contracted logistics or on foreign assistance to supply resources, the first step in reform of the defense logistics system will be to determine which elements of the system the defense sector executes directly and which elements are contracted out. Thereafter, programming may target building the capacity of the defense sector to gain (or regain) direct control over specific elements of the system.

When developing programming for logistical integration, it is critical to develop a contextually driven system that functions well within the confines of the host nation's culture and technical capacity. Programmers must take care to avoid the trap of isomorphic mimicry—that is, when donors transfer their nation's methods of conducting logistical operations to the host nation instead of contextualizing the logistical system to the country's language and technical capacity. For example, U.S. Army logistics systems are all computerized. An army mechanic, after inspecting a malfunctioning vehicle, orders a replacement part through an online U.S. logistical system, which is tied to the higher echelons of supply, ensuring that the ordered part is eventually transferred to the tactical unit requiring it while ordering a replacement part for the supply depot. Placing the order in this way requires a degree of computer literacy and training. Unfortunately, while helping the ISF build an integrated logistical capacity, U.S. forces initially tried to mirror image this same type of system on the ISF. The computer systems provided were in English, which a majority of the ISF personnel did not speak or read; required a constant power supply and continuous access to the internet, which the ISF did not have; and depended on an organizational architecture linked from bottom to top to ensure orders were received, processed, and then delivered—needless to say, the ISF lacked such an architecture.[29] Off-the-shelf solutions will seem easier to implement because they already exist and programmers will have seen them work in the programmers' own contexts, but these tend to be higher-technology solutions that will require substantial levels of existing infrastructure and human capacity to implement effectively. Sometimes lower-tech or even no-tech (e.g., analogue or paper-based systems) solutions may be the better option because they have a higher likelihood of being used and sustained.

CHAPTER EIGHT

GOAL 6
Defense Planning

The sixth goal of defense sector reform is generating effective processes, mechanisms, and systems to conduct strategic and operational *defense planning*. Defense planning can be defined as the "deliberate process of planning a nation's future forces, force postures, and force capabilities (as distinct from operations planning on how to employ the forces in war)" in the near, medium, and long term.[1] It is conducted within the framework of defense policy and in accordance with national security objectives, and translates "security objectives . . . into security and [defense] capabilities."[2] Planning is vital to the efficacy of any defense sector, regardless of size or wealth, and determines the way in which the limited defense resources of the state (human, financial, and materiel) can be effectively employed to achieve national defense goals. Put simply, strategy and policy determine the *what* and *why* of a country's defense goals, whereas defense planning determines the detailed roadmap for *how* defense sector resources can be utilized to achieve those goals.

Defense planning can be broadly divided into (1) strategic planning and (2) operational planning, though the two often necessarily intertwine.[3] *Strategic planning* determines how the defense sector is going to shape the force in the midterm to maintain readiness and in the longer term to maintain the force's military advantage, while operational planning demonstrates how the defense sector is going to employ the force it currently has. Strategic planning in a military environment primarily addresses the medium- to long-term goals of the state.[4] Although it includes operational components, strategic planning also incorporates aspects of planning for the future that address bigger-picture shifts in the nature of the force. It addresses the development of the current and future force, determines what capabilities are still missing that need to be developed to achieve a strategic defense goal, and provides a roadmap for how to develop those capabilities in the midterm and the long term.

Operational planning primarily deals with immediate operations to address actual or potential threats and consequently tends to focus on the short term. Operational planning is conducted by commanders (with significant input from their staff) and is based on existing military doctrine to ensure a unified approach across the defense sector. For example, an operational commander will be given guidance (strategic objectives) about what needs to happen in a given theater and will conduct operational planning to delineate the employment of the force through campaigns that include specific actions (or tasks) that need to be orchestrated in theater within a specific period.

Regardless of the time frame, both operational and strategic plans should be updated regularly. In most advanced Western defense sectors, for example, defense plans are updated annually.[5]

Defense planning tends to be demand-driven, and it relies on other critical defense sector functions addressed separately under goals for strategy generation, financial management, and functioning logistics. Defense planning translates the national security strategy into operational tasks.[6] To that end, the defense sector must have an effective strategy generation process in place, which is addressed separately in this guide under *Goal 9: Strategy Generation*. The relationship between planning and strategy involves a good deal of overlap and interaction but the two terms are often mistakenly conflated.[7] Whereas the defense sector uses strategy to decide *where* and *what* the defense forces will do to maintain national security or international influence, planning is used to determine *how* to use the defense sector's resources to achieve the decisions that were determined in the strategy.

Defense planning is also intimately tied to resources.[8] Planning links the broader strategic goals of the defense sector to the funding and capabilities that are necessary to conduct the missions required to achieve those strategic goals. In all defense establishments, resources (fiscal, human, and materiel) are limited, and effective strategy takes such limitations into account so that realistic planning is possible to achieve the defense goals. Resource-driven planning is based on the available or possibly available resources of the defense establishment and it relies on the defense sector having a system in place to track what resources it has available, including their quantity, quality, and location. Having a "well-functioning resource management system . . . is just as important as having realistic plans for the force structure; in fact, it is inextricably connected to plans which can be effectively implemented."[9] To that end, defense planning must be conducted in tandem with financial planning, which is addressed separately under *Goal 7: Financial Management*.

Defense planning also relies on effective defense procurement and control systems to guarantee the efficient use of resources in line with the planned budget.[10] It grounds strategic goals by identifying which resources and forces will be required to conduct each task within a mission and how to procure them. For example, defense planning determines how many ground forces are needed to achieve a particular defense strategy and why, what capabilities and resources they will need, and how to ensure those forces have what they need at the right time and place. All of this relies heavily on functioning logistical systems, a topic that is addressed separately under *Goal 5: Functioning Logistics.*

Defense planning is a critical component of any effective defense sector; if plans do not accurately capture the defense sector's implementation capacity, the national strategy may not come to fruition. In the absence of effective defense planning, a defense establishment is likely to use resources inefficiently and put lives at unnecessary risk. Transparency and accountability are thus crucial components of defense planning. Although the plans themselves may be classified, the defense sector must be accountable for how resources are being used and how its strategic and operational plans will accomplish military requirements. Transparency is also important when justifying why and how resources being requested will be used and for the analysis and scrutiny of outcomes by the executive, the legislature, and the public at large.[11] To demonstrate the integrity of the defense sector in spending public funds in a responsible and planned manner, it is important within any democracy that defense planning systems include mechanisms for the planned use of resources to be monitored with accountability measures.[12]

Within the defense sector reform literature, there is broad agreement about the importance of effective planning systems to functioning defense sectors and significant discussion of the importance of planning for both the host country and the donor during defense reform efforts. The U.S. Department of Defense DIB Directive, for example, specifies planning alongside policy and strategy as one line of effort within its aim of "creating or improving the principal functions and duties of effective defense institutions," and the OECD DAC Handbook notes that "planning capacities need to be developed and supported at the various levels of government."[13] However, the literature does not provide significant guidance for *how* to establish effective defense planning systems and capabilities within host-nation defense sectors. The UN Defense Sector Reform Policy, for instance, provides guidance for UN planners and focuses on the defense sector reform plan being implemented by the United Nations, but does not specify how to build planning capacity for recipient countries to conduct their own defense planning. NATO's PAP-DIB calls for the "the development of effective and transparent financial, planning, and resource allocation procedures in the [defense] area,"[14] and "a sound national system for planning and managing [defense] and security co-operation, including: appropriate domestic structures and effective working procedures, at both political-military and military level; political guidance and supervision; allotting necessary financial resources," but again does not detail what specific interventions would lead to effective planning capacity.[15]

Guiding Principles for the Design and Implementation of Goal 6

The concept of defense planning is a critical goal for any defense sector institution building and reform effort; however, clear guidance for how to create or improve defense planning across a range of polities is not well defined. The guiding principles below detail two specific baseline elements that will be necessary to achieve the goal of effective defense planning across a wide range of contexts.

First, a defense planning system must direct the efficient identification, coordination, and sequencing of functions and resources to translate strategic goals into operations in the near, medium, and long term. Given the importance of planning to defense outcomes and the complexity of effectively coordinating each element and level of the defense sector to achieve those outcomes, the defense sector must have an organized framework of processes and procedures in place to systematize and coordinate planning processes. "Planning is about understanding how the different variables in a complex adaptive system can interact and posturing the force to exploit, defend, or mitigate the resultant outcomes. *The processes used must account for this dynamism.* [emphasis added]"[16] Processes should be developed to ensure plans are updated on a rolling basis and revised in line with evolving defense realities, and procedures should be developed to ensure that financial planning and budgeting are done in tandem with defense planning for anticipated procurement needs and expenditures.[17]

There is no one system for defense planning. Defense planning takes place across the many tiers and disciplines within a defense establishment, and the planning system or systems in place can vary considerably in their complexity depending on the size and reach of the defense institution. An advanced and complex defense institution such as NATO, for instance, has 14 "planning domains" within its defense planning process that integrate capabilities across NATO member states, including air and missile defense planning, aviation planning, armaments planning, civil emergency planning, force planning, nuclear deterrence planning, and standardization and interoperability planning, among others.[18] The level of complexity of what is being planned is matched by the breadth of involvement in the planning process. The institutions involved in defense planning go beyond just the defense sector and can include "the legislative body with its [specialized] commissions, the government with the key ministries involved in planning, finance and resource allocation, the ministry of [defense], as a stand-alone actor, with its main departments and the military."[19]

Effective defense establishments will have planning systems in place to institutionalize the planning process and ensure that plans are updated when conditions on the ground shift. In most established democracies, this process is often divided into three chronological horizons: (1) short- or near-term planning; (2) intermediate or medium-term planning; and (3) long-term planning.[20] Across different countries, the time periods corresponding to these phases will vary, but in general short-term planning covers the period from six months to two years into the future, medium-term planning addresses the period three to five years into the future, and long-term planning looks five to 20 years ahead. Regardless of the specific time frames assigned to each, these three terms amount to a single unified defense planning system that is inextricably linked to the financial planning of the state and of the defense sector.[21]

Defense planning in the short term is largely bound by the current force status, which is to say that short-term plans are based on what resources and capacity are currently available. Medium- and long-term plans, in contrast, identify what resources and capabilities will be needed in the future to conduct missions based on likely future threats and will include a clear path to acquiring those resources or capabilities. Medium-term planning is required for adjustments to the current force, such as the modification of training methods, changes to recruitment, or the procurement of certain military equipment.[22] Long-term planning will address the structure, size, and posture of the future force as well as the introduction and withdrawal of weapons systems.[23] Having short-, medium-, and long-term planning systems in place allows defense establishments to simultaneously and continuously address current threats; ensure the force is able to defend against rising threats; and make the institutional changes necessary to prepare the entire defense establishment, including the structure of the force and the types of technology it has and knows how to use, to be ready to address future technological and geopolitical developments.

In many countries where defense sector reform will likely be a priority, defense planning systems are often limited, ineffective, or nonexistent. Defense sectors may lack the ability to plan for current threats—either because they lack a system for conducting this planning or lack information about what resources and capabilities are available for conducting effective planning in the near term. Where near-term planning is limited, medium- and long-term planning are also likely absent. Often, there is no formal system for planning, allocating, executing, and accounting for defense expenditures. Where these exist, they are often largely uncoordinated. As a result, acquisitions, payroll, or funding for operations occur largely in a vacuum, separate from planning and budgeting. Defense sectors that lack a system connecting financial realities to the planning process are likely to create plans that are unrealistic and therefore ineffective. Where such systems do exist, defense sectors often struggle to maintain a continuous planning cycle that can adapt and accommodate plans to shifting variables in the near, medium, and long terms.

Still other challenges emerge in countries that have been the recipients of past donor support and that have, with this support, developed medium- and long-term plans. The challenge here is that they have done so only once, and with outside support. Absent this support, there is no system in place for the iterative and ongoing planning required for effective defense sectors. Without a system in place to ensure necessary updates and revisions, such plans are often quickly outdated. Although these countries may have longer-term strategies and strategic goals, they lack the ability to translate these into planning because they have no system for doing so.

In still other circumstances, countries may lack the resources, time, or ability to conduct longer-term planning because they face urgent threats. In such circumstances, longer-term planning may be seen as a luxury they can ill afford. What planning capacity they have is focused on supporting forces in the field or those imminently deploying.

Second, defense planners must have the right skills to create defense plans and to adapt those plans according to changing conditions. Planning is both an art and a science. Defense planning skills must be learned, developed, practiced, and honed to allow defense planners to "[understand] a situation, [envision] a desired future, and [lay] out effective ways of bringing that future about."[24] Planners have to be able to create plans that are measurable in terms of specifications of the final product, achievable within the limits of available human and material resources, and set within a finite time frame.[25]

The range of skills that defense planners are trained to employ throughout the defense planning process includes (but is not limited to) identifying and synthesizing defense objectives, missions, guidance, and ambitions; analyzing relevant actors and the current operational environments in which these missions will be carried out, taking environmental, political, and societal dimensions into consideration; conducting threat assessments; developing an operational approach; selecting courses of action that are sustainable, feasible, acceptable, exclusive, and complete; breaking down activities into tasks; defining the capabilities and resources needed to accomplish the tasks; assessing risks (e.g., from armed actors and from gaps in intelligence) and developing control measures; analyzing what force structure will be appropriate for all anticipated missions and scenarios; performing continuous assessment and incorporating feedback; and adapting plans according to evolving ground realities.[26] Planners also need to understand the financial aspects of each planning decision and be able to adjust their plans within the limitations of the defense budget.[27]

Long-term plans will require planners to be able to identify what is needed within the longer time frames associated with the strategy, such as the development of weapons systems and infrastructural investments.[28] Long-term defense planning requires the planner to have the capacity to analyze trends in the evolution of the defense environment—such as threats that have not yet materialized, technological advancements, or shifting alliance politics—to foresee future defense requirements.[29] Planning for the midterm may require planners to identify the training modifications the force will need to make in order to address an identified possible threat; planners will also have to create a blueprint for how to get the force trained within a time frame of three to five years.[30] For this, planners need to know how to carry out objective analysis and project possible scenarios. Short-term planning will require planners to adapt plans once they are underway and subject to changing dynamics on the ground. As the saying goes, no plan survives contact with the enemy. Planners need strong critical thinking skills as well as the flexibility and attention to detail to make changes as needed. In all three time frames, defense planners will need to know how to liaise frequently with the related, cross-sectoral legislative and governmental organizations and budget planners.[31]

In a democratically controlled military, defense planning relies on civilian and military officials working together to assess, anticipate, and respond to national security concerns. In advanced defense sectors, planning is conducted by both civilian and uniformed generalist planners and specialist planners who have been trained in planning concepts and skills, military doctrine, and planning processes and procedures. Importantly, these defense planners often have years, or decades, of experience in developing defense plans. They will also have had the time to hone their planning abilities and learn to account for changing dynamics, which is important for adapting and updating existing plans, often within extremely tight time frames and under pressure. In countries where defense sector reform will be a priority, however, defense planners are not likely to have this level of training or experience.

Many countries on the lower end of the defense sector capacity spectrum will not have a dedicated cadre of defense planners or specialized training for personnel who are assigned this task. In transitioning democracies, planning, like many other decision-based components of government, may have been purposefully conducted at the higher echelons of government, with little input from subordinate staffs, reducing "planning" to the execution of orders. In the former Warsaw Pact states, for example, defense planning was centralized in Warsaw and Moscow, away from even the capitals of the member states. When the seven non-Soviet states of the Warsaw Pact later began the process of joining NATO, their defense personnel had very little defense planning experience or capacity.[32] In yet other contexts, smart people may be conducting planning, but doing so in the absence of formal training and outside of a planning system, so the potential to develop good defense planners and coordinate their plans may not be realized.

Applying the Goal 6 Principles in Practice

These two principles provide the basis for what defense sector reform practitioners can expect to see across a broad variety of democratic defense sectors if the goal of effective defense planning has been achieved. How the principles are applied in practice, however, will be largely determined by the context of the country undergoing reform. This section provides several examples under both guiding principles to help demonstrate how the application of reform interventions may vary across different defense sector reform environments and to help practitioners think through how to translate the principles into possible activities depending on the status of the states they are assisting.

1. A planning system directs the efficient identification, coordination, and sequencing of functions and resources to translate strategic goals into operations in the near term, midterm, and long term.

The complexity involved in coordinating the defense sector's institutions, its personnel, and its assets to translate strategic goals into appropriate missions, at multiple tempos across multiple time frames, requires an organized *defense planning system*. While there is no single model for systematizing defense planning, the absence of formalized planning systems can lead to plans that are ad hoc, inflexible despite changing realities on the ground, risky, and divorced from the reality of available resources. Across the spectrum of defense sector environments where reform may be a priority, defense planning systems range from extremely complex systems in well-established and -resourced democracies to nonexistent systems in postconflict contexts.

Georgia, for example, has a relatively well-defined defense planning system in place. The Law of Georgia on Defense Planning defines the defense planning system and specifies that state-level defense planning is organized according to the system laid out in the National Security Concept of Georgia and is conducted based on the various parliamentary approved strategies of the state, including not only the National Security Concept of Georgia but also the Threat Assessment Document of Georgia and the National Military Strategy of Georgia. The law also specifies which parts of government have authority over each element of defense planning, including in emergency circumstances. Article Five, for instance, specifies that the planning authorities "shall remain the same during a crisis situation, a state of emergency and/or martial law, and other contingency situations."[33]

The planning system in Georgia is defined at the "strategic" and "agency" levels. The strategic documents listed above (the National Security Concept of Georgia, the Threat Assessment Document of Georgia, and the National Military Strategy of Georgia) make up the country's strategic-level planning, which is conducted in line with the laws regarding each of the three strategies and includes clear lines of civilian authority (the parliament, the government, or the president) for the approval of each strategy.[34] Agency-level planning deals specifically with planning within the Ministry of Defense, and the Law of Georgia on Defense Planning specifies the process for various components of the defense planning system, including the Defense Planning Manual, Major Military Development Programs, Annual Programs, operational plans, concepts, doctrines, regulations, and Defense Planning Regulations.[35] For example, it notes that the minister for defense issues the Defense Planning Regulations that determine the "details and the procedures of agency-level defense planning (terms of reference and deadlines)," and that the Ministry of Defense is responsible for developing the Defense Planning Manual, which lays out the "priority directions and planned activities established by the Ministry for Defense of Georgia for the state defense in the Armed Forces for a multi-year period."[36]

Each of the planning procedures in the Law of Georgia on Defense Planning informs a different component of the planning process. For example, "planned activities aimed to modernize, equip, and train the Armed Forces, to maintain their subunits, to provide logistical support, to maintain and improve the reserve infrastructure, as well as to ensure the implementation of Georgia-NATO cooperation plans" fall under the auspices of Major Military Development Programs. The law also notes that following the last stage of the planning process—the development of Annual Programs—the Ministry of Defense should determine the defense budget based on the plans for existing and new programs. To this end, Georgia's defense planning system incorporates financial planning and promotes efficient use of resources in line with budget allocations: "The priorities of existing [programs] and their funding shall be determined after the inter-agency review of the Annual [Programs]. The [defense] budget shall cover the threats and challenges facing the country, the priority directions established in the military field, and the requirements to ensure [defense] with maximum efficiency. The allocations from the State Budget for the Ministry for [Defense] of Georgia shall be spent in accordance with the annual budget [programs]."[37]

In contexts such as Georgia, where significant donor-assisted reform has already taken place and the defense planning system is therefore relatively well established, defense sector reform efforts might focus on longer-term or ongoing efforts to shape particular aspects of the planning process. In Georgia's case, defense planning systems are relatively robust on paper, but in practice, Georgia struggles to conduct longer-term strategic planning. This difficulty was initially due to the scarcity of resources during the economic crises that followed Georgia's secession from the Soviet Union in 1991, as well as to the absence of a national strategy. More recently, long-term planning has been impaired by the absence of other planning mechanisms. When the Ministry of Defense tried to start a long-term planning cycle in 2006, for example, it was impeded by the absence of national financial planning mechanisms.[38] A sound estimate of what funding the Ministry of Defense can expect over a number of years is essential to realistic long-term defense planning.[39] The ministry did conduct a Development Strategy and Strategic Defense Review that included four-year implementation plans for issues such as force planning and manning priorities, which were linked to recruitment, assignment and distribution, and professional development and promotion policy and plans.[40] However, in reality, the challenge of long-term planning remains, especially planning for infrastructure projects and long-term life cycle planning for equipment.[41]

Unlike Georgia, which has benefited from a significant amount of NATO support and assistance, many states that have recently undergone a democratic transition—such as Tunisia—will not have robust defense planning systems. The challenge in countries like Tunisia is that there is little if any information publicly available about functions that take place within the defense ministry, including defense planning. In some cases, this may be because planning processes and outcomes were considered classified under the previous regime. In others, there is no information because no such formal system exists. There may be informal working systems for near-term planning but no medium- or long-term planning system or capacity. Defense programmers will need to undertake a scoping, mapping, or needs assessment to understand what exists and how planning is being conducted before developing any programming to improve the planning process. A good entry point may be donor personnel who work with recipient defense sector personnel to plan for defense sector assistance, such as the provision of equipment or the coordination of joint operations. Often, sustainment plans for that assistance will provide an introduction to the recipient country's existing defense planning systems and capacity.

In yet other contexts, planning takes place through informal systems that remain mired in residual normative practices from old regimes. In Iraq, for example, there is little to no interagency coordination or attempts by the defense sector to bring other agencies into the planning process. The majority of planning time is spent on deciding how each unit will execute its mission to best accomplish the commander's plan, with little to no coordination with neighboring units. Although U.S. and coalition advisers worked closely with the ISF to develop better synchronized planning systems, the ISF tends to fall back on a course of action dictated by a general officer. There is little, if any, coordination among staff members or between units. This stems in part from the practice whereby major cities are "owned" by different units of the ISF. The Eleventh Iraqi Army (IA) division, for example, was responsible for Sadr City in Baghdad, and any operations that took place in Sadr City were the responsibility of the Eleventh IA alone. Although the Eleventh IA fell under the command of the Rusafa Area Command (RAC), the RAC never coordinated other ISF forces to enter Sadr City. All operations planned for Sadr City were the responsibility of the Eleventh IA commander and his staff.[42] In such instances, defense sector reform may need to focus on generating a formal planning system to replace informal systems tied to individual commanders and on developing mechanisms for coordinated planning. Given the size of the potential reform task, programmers may need to make decisions about what planning functions or aspects to prioritize first while ensuring that any assistance will support the development, over the longer term, of an integrated and coordinated planning system.

Finally, in a postconflict environment, planning systems may not exist at all. For example, Libya's defense sector largely collapsed following the revolution in 2011 and the subsequent internal conflict that shifted security and defense structures toward informal, nonstate, geographically divided entities like militias. While defense planning systems may have existed under Gaddafi's regime, these systems were likely informal, tightly controlled, and largely siloed. In such instances, programming for defense planning systems will be largely foundational, requiring the building of a defense planning system from scratch, including constructing a coordination mechanism with the relevant financial bodies of the new government to establish the budget and align planning with realistic resource availability.

2. Defense planners have the right skills to create defense plans and to adapt those plans according to changing conditions.

Having personnel with the skills to conduct defense planning is an essential but often overlooked component of establishing effective defense planning. In civilian terms, planners are most closely related to project managers, and while some planners are specialists (addressing specific areas of defense planning such as nuclear weapons), all planners require certain skills to create effective plans.

In the case of more advanced democracies, defense planners are likely to have undergone standardized training in skills that are important to the planning process, such as synthesis of strategic guidance, analysis of mission operational environments, assessment of threats and risks, development of task sets to achieve activities, identification of the capabilities and resources needed to accomplish the tasks, analysis of appropriate force structure, monitoring and assessment, adaptation of plans in real time, and comprehension of the financial aspects of each planning decision.

Planning skills are not innate. To become effective planning team members, planners require education, mentorship, and practice. In advanced democracies, schools have been developed to teach the science and art of planning. These schools, such as the U.S. Army's School of Advanced Military Studies (SAMS) or the Air Force's School of Advanced Air and Space Studies, focus on teaching the next generation of the military's planners, strategists, and leaders. The Advanced Military Studies Program at SAMS, for instance, lists developing "effective planners who help senior leaders understand the operational environment and then visualize and describe viable solutions to operational problems" among its key goals.[43] The curriculum is focused on a multidisciplinary approach to training planners to understand problems, develop viable options, and resource potential solutions to complex problems. Countries that need to improve their defense planning efficacy must devote attention and resources to the professional education of their planners. Effective planning requires a specialized skill set not often taught in more conventional military education programs.

In countries where defense sector reform is likely to be a priority, those assigned to conduct planning often receive little to no extra education on how to conduct this critical function. The Iraqi military, for example, relies heavily on senior grade officers who have no planning education. In such cases, defense sector reform programmers may need to identify interim steps to gradually institutionalize planning training into the existing training process. For example, training may be initially developed to create a cadre of specialists with planning expertise to lead the joint planning process, while planning skills are gradually introduced throughout the defense sector's existing training structure. In the case of Iraq, the United States International Military Education and Training Program provides Iraqi officers with the opportunity to attend U.S. war colleges and command and general staff colleges, so that ISF members can acquire a better understanding of how other militaries plan and how the ISF can better develop its own planning processes.[44]

In such contexts, informal cultural norms may also stop planners from coordinating across and between departments and agencies. In the case of Iraq, military planning within the ISF is highly stove-piped, and the efficiency of that planning depends on the unit or level of advisers available. Unlike the U.S. military, which relies on the professionalism and expertise of its NCOs, ISF planning efforts are undertaken only at the officer level and often only at the general officer level. In contexts where a culture of noncoordination and sparse communication is embedded in the system, defense planners will need to be trained to coordinate, communicate, and plan jointly. A possible entry point may be to develop joint training for commanders at the same level who conduct planning in different units, agencies, or departments to initiate communication and demonstrate the importance and value of joint planning. Often, in such contexts, there is a need not only to develop planning skills but also to effect a change in the military's mindset about how and why information should flow and why planning must be coordinated.

In postauthoritarian countries where the military establishment was previously viewed as a potential threat to the regime, planning skills are likely to have been tightly controlled and planning staff would have focused on executing orders, not developing plans. This was the case in Libya before the revolution, where Gaddafi, along with a few handpicked people at the top, dictated the defense plans to the defense sector personnel, who were there to execute those plans without input or question. In postauthoritarian countries such as Libya, once the conflict is stabilized and defense sector reform activities can begin to rebuild the defense sector, establishing defense planning skills will be an important function that will likely have to be generated from the ground up. This may require introducing basic planning skills in the near term to prepare personnel to implement newly developed defense sector planning processes. As these processes are developed and gradually institutionalized, longer-term programming could aim to develop a specialized cadre of planners for the defense sector.

Countries that have specialized, well-trained defense planners may benefit from specialized planning assistance and training. In Georgia, the responsibility for planning is assigned to the Ministry of Defense's Civilian Office, but strategic-operational planning is also conducted by the general staff, which has a joint planning structure: J1 for organizational planning and personnel management, J2 for intelligence planning, J3 for operational planning, J4 for logistics planning and management, J5 for strategic planning, J6 for communication/signal management, and J7 for planning of military education and combat trainings.[45] Despite its robust planning system, Georgia's planners still struggle to conduct and implement long-term or strategic planning. Programming could be developed to train planners to conduct effective strategic planning, such as building planners' capacity to analyze trends in the evolution of the defense environment to better account for future defense requirements.

CHAPTER NINE

GOAL 7
Financial Management

The seventh goal of defense sector reform is the effective and transparent financial management of the defense budget and spending to enable the defense sector to perform its mission as defined in national policy and strategy. In democratic systems, defense funds are public funds, and the allocation and expenditure of those funds must be transparent and subject to public scrutiny. Although the mechanisms by which this is accomplished vary widely, individuals with the power of the pen and the purse must ultimately be accountable to the citizens they serve. Rules and procedures further safeguard the expenditure of public monies, girded by accountability mechanisms, anticorruption procedures, and audits. An effective and transparent system for financial management of the defense budget and defense spending is one that can plan for, allocate, execute, and account for resources expended for national security; accurately budget short- and long-term requirements; allocate resources for strategic priorities; execute budgets; oversee contracts and acquisitions; and report on and audit those expenditures. The ultimate measure of an effective financial system for the defense sector is the degree to which it can financially plan for and provision the achievement of objectives and priorities in the national strategy.

In large and well-resourced democracies, complex systems have evolved to manage the expenditure of public funds for national security and defense. For example, the United States uses a planning, programming, and budgeting execution system to allocate resources annually in line with strategic objectives,[1] and many NATO members have adopted similar systems.[2] Smaller, less well-resourced democratic states may not have as complex a system for managing defense sector finances, but their systems still feature transparent access to information about how public monies are expended. These processes and the annual presentation of budgets and review by legislative bodies are transparent. Citizens can, to a certain level of detail, know how funds are being allocated and spent, and why.

In much smaller, poorly resourced, or transitioning authoritarian states, which are often the most likely candidates for defense sector reform efforts, what constitutes the system for managing and spending defense funds is often best understood by what is missing. At the macro level, systems for planning, allocating, executing, and accounting for defense expenditures are often inchoate. In place of a system, there are largely uncoordinated processes for budgeting, planning, and procurement such that acquisitions, payroll, or funding for operations occur largely in a vacuum, separate from planning and budgeting. In such systems, forces are not paid or are paid irregularly; the acquisition of new weapons systems is approved but never executed; and forces are deployed without provisions or arms. In such instances, there can be no external transparency because there is no internal transparency; senior defense officials do not know their own defense budgets or how or why money is allocated across the defense organization.

In other instances, public budgets capture only a small portion of defense sector funds, the majority of which are managed off-budget, set aside for government capture industries, or lost to corruption. In resource-scarce environments, published budgets often bear no relation to reality; what the defense organization is slated to achieve on paper does not match the actual funds it can spend. In such instances, there may be a system for financial management of the defense budget and spending, but it is ineffective or irrelevant.

At the micro level, there are equally concerning gaps. Financial planners may have no planning skills. Accounting staff may lack basic knowledge about accounting practices, including how to prepare a balance sheet, track expenditures on a budget, or plan for and provision short- and long-term needs. Budget and finance offices may be staffed by individuals who lack even basic skills such as numeracy.

Regardless of the type of gaps or the reason for those gaps, countries that lack effective systems for managing defense budgets and spending struggle to meet their national defense goals and priorities or, in the absence of a strategy-generation process, their defense sector missions and mandates, with serious consequences for military effectiveness. It is thus not surprising that the defense sector reform literature devotes a great deal of attention to financial planning, resource allocation, defense spending, and legislative oversight.[3] However, practice does not always follow theory: "Experience has indicated that issues of fiscal sustainability tend to be ignored, and standard public finance management approaches are rarely included in SSR programmes."[4] The main challenge in tackling this problem often lies in the unwillingness of a government to "open its coffers" and reveal what is frequently guarded as essentially a state secret—the size of the budget and how, and on whom or what, it is spent. Such a disclosure may threaten the source of power or wealth of key individuals. If budget and financial details of spending and procurements are closely guarded, then reforming the financial management *process* may be a starting point. But this requires understanding what processes currently exist and how decisions are made, tasks that are made more difficult when key defense staff do not know how their own system works.

Guiding Principles for the Design and Implementation of Goal 7

Given the centrality of effective and transparent financial management for achieving the mission and strategic priorities of the defense sector, it is an essential goal of defense sector reform. Transparency in defense sector decision-making is addressed under the defense sector goals of democratic control of the defense sector (Goal 1), civilian control (Goal 2), and legislative and judicial oversight (Goal 3). Here, the focus is on the transparent financial management of the defense budget and spending. The following three principles offer guidance for achieving that in a range of likely environments for future defense sector reform.

First, there must be a system or, at a minimum, coordinated procedures for planning for, allocating, executing, and accounting for resources expended for the defense sector. This need not require an overly complex, automated, or high-tech system. At its most basic level, such a system should ensure that defense sector officials know the amount and type of funds they are authorized to spend in a given budgetary cycle and what likely funds will be available in future budgetary cycles such that they can plan expenditures in line with defense sector needs and priorities, including life-cycle costs for new acquisitions. Such a system should also enable the execution of expenditures. At a minimum, defense sectors must be able to pay and equip military personnel and fund core operations. Finally, there must be some accountability for how funds are expended, which requires records of those expenditures to which defense sector officials, relevant oversight authorities, and citizens have access.

In many of the environments where defense sector reform will likely be a priority, defense sector officials lack access to this most basic information, either because it simply is not tracked or recorded or because it is closely guarded. In transitioning authoritarian countries, finance and budget staff may have only limited authority to execute transactions unless they receive orders and often have no insight into the defense sector's wider finances. In other cases, the problems stem from constraints on state revenue; the defense sector has little visibility into its budget because finance ministries lack the capacity to track, account for, or plan government revenue and expenditures.

Systems for planning and systems for expenditures may exist but are un-coordinated. Alternatively, systems for coordinated planning and expenditures may exist but lack capacity for longer-term planning, including the life-cycle costs of new acquisitions. Because planning, allocation, execution, and record keeping are systemic functions, a breakdown at any point in the system can render the entire process ineffective. And when defense sector officials them-selves have limited or no access to information, either about their own budgets or about the larger system, then oversight and transparency are likely also ab-sent. Indeed, herein lies one answer to the question of return on investment for security assistance programming. Often, donor countries provide equipment to host countries without attention to how that equipment can be sustained. Where sustainment is considered, attention is often paid to training personnel how to use or repair the equipment. What is frequently overlooked is how the recipient country can sustain the costs of that new equipment, including personnel, training, spare parts, and repair costs in the outyears. Sustaining the costs of new equipment can be difficult to do if the recipient defense sector does not itself have a system for short- and long-term financial planning.

In most instances, a likely starting point for defense sector reform is iden-tifying and mapping what currently exists and understanding how it functions and why. Such an approach guards against importing a system from elsewhere that may not address the underlying causes of the lack of efficiency or capac-ity. A high-tech system will likely fail if the institution lacks the appropriate infrastructure or sufficient staff with basic computer skills; a system with tiered decision-making authority will fail to deliver results if there are no standard operating procedures or guidelines that authorize staff to make those decisions or if there is no training for staff on accounting and financial planning best practices. Systems are inanimate; people must embrace them, understand them, have the skills to use them, and commit to their implementation if those sys-tems are to have their intended impact.

Second, there must be clearly articulated rules and procedures for financial management decisions, defense sector acquisitions and procurement, and audits. As explained by one leading defense institution building expert, these rules and procedures must comply "with applicable laws and ethical standards" for de-fense sector personnel (both uniformed and civilian); with "governmental and internal accounting policies, including transparency polices for public spend-ing"; and with "internal control arrangements and procedures in line with ac-countability requirements."[5] Normally formalized in accounting policies and manuals of financial procedures, guidelines should also include "procedures for monitoring and recording assets received, held, and spent."[6] Furthermore, procedures should "regulate the development of budgets, the processing of financial assets (i.e. receiving, recording, securing, deposing, and spending processes), the authorization, recording and monitoring of expenditures, the establishment of auditing and controlling authorities and agencies . . . within the defense sector, their remits and operating procedures, and the procedures for contracting, sub-contracting, and outsourcing."[7] For major acquisitions and procurements, this may include establishing "a hierarchical authorization system . . . up to parliaments for high-value assets."[8] In addition, rules and procedures must detail the mechanisms by which to ensure there is transparent oversight of these financial processes. "The financial system should be transparent to the legislators, auditors, governmental regulators, and to the public in order to provide for democratic oversight as well [as] building and maintaining public trust."[9]

In most defense sector reform contexts, there are likely one of two problems: either there are no rules and procedures, or rules and procedures exist but are flawed or not enforced. Absence or poor enforcement reduces oversight, transparency, and the effectiveness of audits, and opens the door to corruption. In such conditions, defense sector personnel may resist defense sector reform efforts or actively work to undermine them. Rules and procedures may complicate the day-to-day work of personnel, increase their workload, or limit their opportunities for personal profit or profit for their patronage networks.

Third, defense sector budgets and budget processes must be transparent and subject to oversight. Defense sector resources are frequently scarce, and decisions about their allocation will be heavily contested. Defense sector personnel may resist exposing budgetary negotiations to public scrutiny, fearing either the loss of efficiency, their ability to negotiate backroom deals or tradeoffs, or the exposure of corrupt practices. Resistance may also come from a lack of understanding of why transparency is essential. Indeed, in most NATO nations, and for a long period of time, the "defense budget was a secret and centralized affair, as the government budget often was."[10] It is thus not surprising that many of the countries that are likely candidates for defense sector reform will view defense budgets and the budgetary process as sensitive, if not secret, and resist exposing this information to view by the public and by existing or potential adversaries. Nonetheless, transparency of budgets and budget processes is essential because the defense sector has no marketplace competitor. "It will always be only one supplier—the state, who sets the prices, as it will set the quality and quantity of the defense 'product'."[11] Because the defense sector is a monopsony and defense is a "public good," defense sector "managers and financial planners . . . operate in different conditions from the market environment."[12] Those conditions require levels of oversight and transparency that may expose information to the public and reduce efficiency, but in so doing also ensure that public funds are expended appropriately and accountably.

Applying the Goal 7 Principles in Practice

These guiding principles demonstrate what practitioners can expect to see across a broad variety of democratic defense sectors if the defense sector can plan for, allocate, execute, and account for resources expended for national security; accurately budget short- and long-term requirements; allocate resources for strategic priorities; execute budgets; oversee contracts and acquisitions; and report on and audit those expenditures. Applying the principles in practice, however, will vary widely, depending on the local context. The following examples, drawn from various reform environments, help to demonstrate how the application of financial management reform interventions can differ. These cases aim to help practitioners consider how to translate the three principles into possible activities by demonstrating how different each reform context can be and how reform will need to be tailored to the specifics of the country at hand to create an effective and transparent system for financial management of the defense budget and defense spending.

1. There is a system for planning for, allocating, executing, and accounting for resources expended for the defense sector.

The measure of an effective defense sector financial system is the degree to which it can financially plan for and provision the achievement of objectives and priorities in the national strategy. To do so, the defense sector must have a system in place for the planning for, allocating, executing, and accounting of defense resources. Such systems ensure that the amount and type of funds the defense sector is authorized to spend in a given budgetary cycle are known; they also provide a clear indication of what likely funds will be available in future budgetary cycles such that longer-term expenditures (e.g., life-cycle costs of new equipment) can be planned. These systems make possible the payment of personnel, the acquisition of equipment and materiel, the funding of operations, and, importantly, the recording of accurate financial expenditures to ensure accountability. In the environments where defense sector reform will be a priority, the degree to which countries have such systems in place will vary widely.

In relatively advanced democracies, the financial planning system is likely to be fairly well established, though often with legacy issues that have to be overcome from previous, often authoritarian, governments. In Georgia, the Ministry of Defense supports the minister of defense in managing the Georgian

Armed Forces and is responsible for developing defense financial management policies regarding, for example, the management and allocation of resources, planning and conducting of procurements, and planning and execution of the defense budget.[13] In the area of resource management, the Ministry of Defense has made strides since the early 2000s to shift from ad hoc and inconsistent resource usage to longer-term resource planning and management. The reforms have created the appropriate mechanisms for national security and defense needs to be translated into resource-allocated defense plans. The recent planning, programming, and budgeting system institutionalization process ensures defense objectives are followed with relevant and realistic implementation plans, including, importantly, transparent budgeting and procurements plans. The Ministry of Defense submits program budgets to the Ministry of Finance that are detailed and include all program costs.[14]

However, Georgia's financial management system is not without problems. The Ministry of Defense has to constantly provide training for its financial staff, because there is a lack of personnel with sufficient technical skills in financial management or accounting techniques. Frequent changes in the ministry's leadership have also had a negative effect on the financial decision-making process; the ministry is highly centralized and each new minister brings with him or her a new vision for how funds ought to be spent, which in turn disrupts longer-term programming. Another problem is posed by the fact that, given weak parliamentary oversight, the ministry has been able to evade legislative scrutiny, especially on confidential contracting, such as the controversial acquisition of an air defense system from France in 2015.[15]

In cases such as Georgia, defense sector reform practitioners will be faced with financial management systems that are established but have specific areas that require significant reforms for the overarching planning, allocation, execution, and accounting functions to work effectively. Georgia's lack of skilled financial personnel, for example, is likely to be a common problem in defense sector reform environments. In some cases, the problem will be with recruiting the right personnel, addressed separately in the guide under *Goal 8: The Right People*. In many countries, the defense sector cannot compete with the salaries being offered in the private sector for areas requiring special skill and knowledge sets. In such cases, the solution may be to consider programming to help the country to improve recruitment and retention or benefits, where possible. In other cases, years of war may have denied all but a handful of people the opportunity to develop financial skills. In countries where personnel have the capacity but not the necessary training and education, programming may focus instead on providing either curriculum development in such skills as accounting or financial planning or giving small cadres the opportunity to attend courses in the donor country where they can learn these skills and bring them home to teach others.

Similarly, Colombia's defense sector has clear policies for financial planning and resource allocation. Each military unit defines its needs based on capacity strengthening, institutional transformation, fiscal limitations, and the priority of each project.[16] Each unit also takes into consideration scenario drills and capability projection for four years, although the budget is reviewed and approved every year. Each service prepares a preliminary draft budget on an annual basis that is then reviewed and approved by the commander, the minister of defense, the National Planning Department, and the Department of the Treasury. Once reviewed by the executive, the preliminary draft budget becomes the annual operating investment plan. It is analyzed and approved by the Congress, and it is regulated through the budget bill.[17]

Despite a robust financial planning system, Colombia has to contend with the fact that its defense budget is reliant on the fluctuating economy. In general terms, Colombia has the capacity to finance its defense sector, even though its defense budget is often among the highest in Latin America. However, this capacity is directly related to the health of the country's economy. The budget started growing in 1998, based on increased national expenditures and on the international resources provided by the United States for Plan Colombia. Defense financial resources also increased through a war tax imposed from 2002 to 2006, and two additional taxes imposed from 2006 to 2014. The end of the special taxes, along with declines in the past decade in the revenue from two of Colombia's main exports—oil and mining products—has resulted in lower production and investment, which have in turn made the defense budget vulnerable to cuts.[18] Compounding this issue, the United States has gradually reduced assistance from approximately US$700 million in 2008 to $529 million in 2017, highlighting the need for Colombia to find ways to ensure adequate and sustainable funding of its defense sector.[19]

Many countries slated for defense sector reform will face similar challenges. In such instances, assistance may be provided to help defense sector personnel better manage the financial resources they are granted in the national budget, including through fiscal planning and cost-saving measures. If programming consistently outpaces available budgets, closer analysis may be required of the defense planning process, which is treated in this guide under *Goal 6: Defense Planning*. If gaps are driven by a lack of access to information or poor coordination, programming can attempt to streamline information sharing or enhance transparency, particularly at the detailed line-item level. Because addressing gaps will require a high degree of technical expertise, programmers may start by sending an expert team to conduct an initial assessment to identify what specifically is missing or what mechanisms could benefit from institutional strengthening to ensure that any programming addresses the underlying issue generating a breakdown or inefficiency in the system for planning for, allocating, executing, and accounting for resources expended in the defense sector.

2. Clearly articulated rules and procedures direct financial decision-making, acquisitions, procurement, and audits.

The legal rules and procedures that govern financial management decisions, defense sector acquisitions and procurement, and audits are critical for oversight, accountability, transparency, regulation, and monitoring of the defense sector's financial assets. In Western democracies, such rules and procedures are clearly articulated and institutionalized both throughout the defense sector and across the broader government institutions that play a role in financial management. In countries where defense sector reform is a priority, such rules and procedures may be flawed, unenforced, or missing.

In more advanced democracies, the rules and procedures for financial management decisions, defense sector acquisitions and procurement, and audits are often well established, if imperfect. In Colombia, for instance, auditing of the military budget falls under the domain of the Office of the General Auditor within the Office of the Comptroller General, Colombia's leading fiscal accountability institution. In addition, each institution has guidelines and protocols to assign and approve budgets, as defined by law, that also apply to the Ministry of Defense. Each institution manages its budget in accordance with national procurement law and other laws concerning budget management.[20] The Internal Control Office of the Ministry of Defense conducts audits for the ministry and the military forces. The president appoints the head of the office to ensure its independence from the institution, and reports issued by the Internal Control Office are published on the office's website.[21]

Regarding acquisitions and procurements, all Colombian government institutions are required to publish annual procurement plans; the defense sector has its own plan, which complies with acquisitions regulations and is published in the government's electronic procurement system. This requirement allows the public to monitor the Ministry of Defense's purchases of goods and services.[22] The ministry and the National Planning Department supervise the purchase of all military equipment, including airplanes, tanks, helicopters, and other heavy equipment. Technical and ethics committees provide oversight of the procurement process. There are also single-source, noncompetitive acquisitions of goods and services that do not require public reporting due to their national security sensibilities.[23]

In Georgia, the Department of Internal Audit in the Ministry of Defense is relatively new, created in 2014 as a component of the European Union–Georgia Association Agreement, which supports public finance reform and monitoring of the defense budget by harmonization with EU standards on the basis of the concept of public internal financial control (PIFC).[24] The Department of Internal Audit is subordinate to the minister of defense and provides assurance and advisory services to the minister. The department's key roles include assessing the effectiveness of the ministry's risk and financial management and other activities, observing the ministry's compliance with Georgian legislation, and issuing recommendations and monitoring their execution. The department conducted a pilot audit in 2015–16 with the support of an international consultant and prepared recommendations.

The Ministry of Defense's State Procurement Department has also made significant progress in making the defense procurement processes more transparent.[25] In the recent past, classified procurement and the issuing of secret contracts were a major source of corruption in Georgia. Not all participating companies within the classified procurement process had access to the tenders' terms and conditions, calling into question the transparency of the decision-making process. In 2013, the State Procurement Department launched a new online portal where it electronically publicizes requests for proposals.[26] The implementation of electronic tenders has helped create a competitive environment for interested qualified firms. It is an open-procurement platform that ensures equal treatment with no bidders getting prior or inside information regarding contracts. Furthermore, it has given representatives of the nongovernmental sector access to monitor the tender processes both online and by participating in tender commission meetings.[27] The percentage of classified contracts gradually decreased to around 10 percent in 2017, although that percentage is still double NATO standards.

Although the institutionalization of an internal audit department and the improvement of the transparency of contracting are critical steps in realizing the second principle, the rules and procedures for financial management do have shortcomings that impact effectiveness. One problem is a basic lack of understanding surrounding the term "audit" and the role of the department within the Ministry of Defense. Despite the order of the minister of defense that clearly defines the department's mission, structure, and responsibilities, colleagues from other departments still tend to perceive the department as punitive and created in order to catch them doing something wrong.[28] Another issue is that because this functional skill set is new in Georgia, there is a significant shortage of professional staff able to conduct the auditing function. Compounding this issue is the problem of retaining qualified and skilled employees, who are often tempted by higher salaries in the private sector. Lastly, subordination to the minister, who is a political figure, instead of falling under the purview of a permanent undersecretary of defense, as is accepted practice in Western countries, creates a risk that internal audits may become a tool for political manipulation.[29]

In countries where rules and regulations for financial management are relatively new, there are likely a host of entry points for defense sector reform programming. The first step will be determining what rules and procedures exist and how well they have been established and institutionalized. Process streaming may also be a useful way of determining what hurdles or gaps need to be addressed if nascent systems have been developed but do not yet function efficiently or effectively. In some instances, a new system may have been provided by a donor, but staff are not well trained in how to use it or adhere to it. In other instances, record keeping or tracking may be weak, requiring a combination of technical and advisory programming to strengthen those functions. Frequently, inherited systems will not support new financial management requirements because the systems are not efficient or do not promote the appropriate degree of transparency. Countries such as Tunisia that have undergone recent democratic transitions will likely feature a range of challenges related to transparency and access to even the most basic information required for effective financial management, acquisitions, and procurement. In these contexts, audits may be a largely new function and require a host of capacity building and technical support measures to implement fully and effectively.

In other contexts, such as Libya, there will be no clear rules or procedures surrounding the defense sector's financial management—and many individuals within the system will be working hard to keep it that way. Amid the chaos created by Libya's civil war and ongoing internal conflict, many individuals have profited from corruption, especially within the defense sector. For example, some officials use their positions to purchase materiel and equipment that is then resold for a profit on behalf of their tribe or a militia with which they are associated. Militias have managed to plant (or buy places for) their members in positions of power within the ministries that allow them to control revenue flows from sources such as ghost soldiers on the payroll, the acquisition and then resale of materiel, "running" detention centers (that do not exist or which they spend no money to run) for which they receive government stipends, collecting "taxes" at border or territorial crossings, and so on. Without clearly defined rules and regulations to deter this corruption, corruption itself has become the system in Libya.

In countries where defense sector personnel can profit—in money or status—from the absence of rules, implementers are likely to meet active resistance to any proposed reform or to the establishment of financial management rules and procedures that could limit opportunities for corruption. In such cases, defense sector reform implementers will need to devote time and energy to ensuring there is sufficient political will to undertake reform. A good entry point may be to devise small improvements that generate "quick wins" to demonstrate the value and benefit of reform. But if corruption is endemic, more fundamental reforms will be required, likely beginning with the legal and regulatory frameworks addressed under *Goal 1: Democratic Control* and transparent oversight mechanisms addressed under *Goal 3: Legislative and Judicial Oversight.* Once these standards and prerogatives are defined in statutes or in a constitution, translating them into clearly articulated rules and procedures for financial decision-making, acquisitions, procurement, and audits will be an important step to take. Such new rules and procedures will likely also require significant capacity building efforts for staff, ensuring they have the right technical expertise to implement and abide by the new standards. If such expertise cannot be generated through training, activities to support recruitment of this technical expertise—treated under *Goal 8: The Right People*—may be required, especially if corruption is truly endemic.

3. Defense sector budgets and budget processes are transparent and subject to oversight.

In democratic defense sectors, the defense budget and budget processes must be transparent and subject to oversight. This ensures that the public funds being spent in the pursuit of the nation's defense are subject to oversight and scrutiny, which are essential to the accountability of the defense sector. While this budget transparency is now the norm in most Western democracies and is often being worked toward in newer established democracies, in many countries where defense sector reform will be a priority, the culture surrounding the defense budget will be one of secrecy. This is likely to have resulted in limited oversight, serious inefficiencies, widespread misuse of funds, or rampant corruption.

In countries that are only recently emerging from conflict, the defense sector is unlikely to have a robust financial system, and the defense sector's leadership may not even know what the defense budget is (often because the government itself has not been able to allocate a budget due to its own failing institutions). In Libya, for example, the conflict that began in 2011 has left the country divided between the UN-backed government, the GNA, in the west and a de facto militia government in the east. The GNA's budget is largely spent on the ongoing internal conflict to the detriment of almost all other public services. That said, spending money on defense and having a defense budget are not the same. In Libya, those within the defense sector are hard pushed to answer questions about their budgets, and while the government itself has made announcements about defense spending—for example, in the wake of reignited conflict in 2019, the GNA announced it had allocated 40 million Libyan dinars (US$28.5 million) to its defense ministry—there is no clear financial planning or data to support such announcements. It is unclear how the budget is allocated or what the annual total is for each department, even to those within the defense ministry.[30]

In cases where the issue lies in the defense sector's budget, the problems can range from countries where the budget is known but is not enough to cover what the defense sector needs, to countries where the defense sector has no idea what its budget actually is, to other instances where the budget that is written down on paper is nowhere close to the budget that the defense sector receives in reality. In cases such as Libya, an assessment is often the first course of action, because without a clear picture of what exists, it is difficult to determine if there is anything that can be salvaged, although the outcome of the assessment is likely to reaffirm that the systems for planning, allocation, execution, and accounting need to be rebuilt from scratch. In countries where the conflict has not yet concluded, the first hurdle for programmers will be to determine at what stage the country will be ready to receive and sustain reform programing that focuses on building such institutional systems.

In more advanced democracies, however, institutionalizing oversight of the defense budget may be well underway. In Colombia, the defense sector has taken several measures to ensure financial transparency and oversight. The Office for Internal Control at the Ministry of National Defense, for instance, oversees anticorruption efforts. Article 29 of Decree 4890 of 2011 stipulates that this office oversees creating and implementing strategies to ensure ethical behavior. Under the Anti-Corruption Statute of 2011 (Article 8), most of the office's staff are civilians rather than military personnel, a provision intended to ensure the staff's independence from the institution they monitor.[31] Several recent entities also focus on improving transparency and accountability within the military. In 2014, the Action Group for Institutional Transparency was established to monitor financial transactions with the Ministry of National Defense and identify those that need additional investigation. In 2016, as part of a restructuring program, the National Army created the Office of the Application of Norms of Transparency in the Army, which operates as a transparency oversight entity within the ministry and works closely with NATO and its partners.[32] Colombia also joined the NATO Building Integrity Program, which provides external oversight of transparency and accountability in the military sector.

However, Colombia's financial transparency system is not without weaknesses. For example, despite the relatively robust regulatory framework, the lack of civilian personnel specialized in defense and security issues remains a problem. This is evident not only in the Ministry of National Defense but also among members of Congress, journalists, and academics, most of whom lack expert knowledge of military affairs.[33] While guidelines and controls for how to invest resources exist, civilians in the ministry and in the National Planning Department who lack the required technical knowledge often allow the purchase of materials and equipment that are well beyond the scope of what a project needs. The dearth of technical knowledge threatens the credibility of the oversight mechanisms and may allow for corrupt accounting, acquisition, or procurement practices to go unnoticed and unchecked.

Many countries share Colombia's challenge: an oversight system exists but those charged with conducting oversight do not have the military knowledge or background to fully understand what they are being asked to review or the financial expertise to detect and investigate irregularities or potential violations of established processes and procedures. In such cases, programmers may take a two-pronged approach to addressing the issue. First, programming could focus on improving the regularity and transparency of communications involving budgetary needs. Second, programming could offer to provide defense finance literacy training to targeted audiences that have financial oversight functions within or outside of the ministry of defense. In both cases, improving the ability of those with oversight responsibilities will be a key, long-term goal for the democratic oversight of the defense sector.

CHAPTER TEN

GOAL 8
The Right People

The eighth goal of defense sector reform concerns what is arguably the defense sector's most important asset: people. To effectively defend the state, the military must have *the right people*, with the right skills, in the right positions, at the right place, and at the right time.[1] Whether it is termed "human resources management," "personnel management," "human capital," or "skills-based management," having the right people in the defense sector generally requires policies, rules, and practices to effectively and strategically plan for, identify, recruit, develop, retain, retire, and manage civilian and military personnel.

In a democracy, the defense sector workforce includes both military and civilian personnel. Civilian personnel support the uniformed military in achieving the country's national defense goals and objectives through their skills in fields such as information technology, intelligence analysis, healthcare, and education, among many others, and include government employees as well as contractors. Military personnel include all active enlisted personnel, commissioned officers and NCOs, and reserves (if the military has reserve troops) of the country's armed forces. Whereas civilians working in the defense sector are subject to personnel management policies much like those seen in the private sector, the armed forces are subject to policies that differ significantly from their civilian counterparts, including terms of service that by law must be fulfilled and military justice systems that are based on military-specific codes of conduct.

The balance between civilian and uniformed personnel varies among democracies, as does the complexity of the systems in place to manage them. In large democracies, the sheer number of people in the military and supporting civilian staff, who are often spread across the globe, as well as the existence of advanced personnel management capabilities, may mean that personnel management systems are complex, involving thousands of subsystems and dependent on high-tech mechanisms. In smaller, less developed states, however, the system may be limited to paper and centrally managed by just one office; despite its lack of sophistication, if it covers the necessary elements of personnel management, the system may be highly effective.

The importance of having the right people is discussed at length in the defense sector reform literature, as are the myriad ways that implementers can potentially help improve defense personnel skills. For some, such as the U.S. Department of Defense, the emphasis is on building a partner country's capacity to conduct human resource management, where the development of individual personnel is separately addressed in programming focused on enhancing professionalism or on defense education and training.[2] NATO also directly lists the development of "effective and transparent personnel structures and practices in the [defense] forces," as one of its key objectives for defense institution building.[3] The OECD emphasizes the importance of developing the skills of security sector personnel not only for the effectiveness of the defense sector but also for the success of the defense sector reform process itself.[4]

The literature also focuses on instilling democratic standards that the people and personnel management processes in the defense sector should embody, such as "civil and personal rights; adherence to organizational norms defending individuals; provision of a reintegration system into the civilian society; guaranteeing healthful working conditions; and providing opportunities for individual development."[5] In countries undergoing defense sector reform, the adoption of such norms may represent a significant cultural shift for the defense workforce. The United Nations and the OECD thus emphasize the importance of teaching and codifying human rights and gender and minority inclusion norms, as well as monitoring processes in defense sector institutions.[6] Such efforts are achieved through education, training, and the implementation of policies that comply with international human rights standards not only for defense sector personnel, but also for any people (e.g., prisoners of war) for whom defense personnel will be responsible.[7]

Guiding Principles for the Design and Implementation of Goal 8

Challenging as defense sector personnel management may be, the effectiveness of the defense sector relies heavily on having the systems, policies, and processes in place to ensure the workforce is made up of the right people to achieve its objectives. To that end, effective personnel management in the defense sector should abide by the following three principles.

First, the defense sector must strategically plan for an effective and appropriate workforce to staff the military and its supporting functions. To carry out its missions and achieve its objectives effectively, the defense sector should plan to have the right mix and number of civilians and military personnel with the right capabilities.[8] The defense sector must have the capacity to analyze current and potential future operating environments to inform decisions about what constitutes the optimum size, composition, and skill sets of defense personnel. For example, at the strategic level, analysis will help determine the right balance of civilian and military personnel, which, along with clearly delineating roles and responsibilities between civilian and military leadership, is necessary for effective democratic control.[9] In many polities, such personnel decisions, together with decisions about the composition and command structure of the armed forces, are also subject to parliamentary scrutiny and approval.[10] This requires that individuals implementing legislative oversight understand the personnel decisions at hand and are given appropriate access to information by the defense sector to do so.[11]

Knowing how many people the military currently employs, what those people are doing, and where they are located is a first critical step to determining what human resources the defense sector has and needs. The availability of accurate and current data about the human capital of the defense sector—ranging from data about the number and location of current employees, to data about recruitment, career progression, training, absenteeism, productivity, and personal development—is critical for developing a realistic defense strategy. It also has a direct impact on the effectiveness of the military in making the most efficient use of its human resources to achieve missions in support of the defense strategy.[12] Further, this data is critical for identifying which defense sector employees do not meet the defined criteria for recruitment or leadership, and making adjustments through additional education and training to fill the skills gaps, restructuring to appropriate levels based on skill, or introducing a process to remove unqualified personnel.

In more advanced countries where such data is reliably tracked and available, the defense sector can make better, data-driven decisions about staffing and budget allocations, make human resource processes more efficient to improve performance, plan better for future staffing requirements, and react faster with more accurate decisions about staffing when the situation on the ground rapidly changes. In countries where defense sector reform is likely to be implemented, however, personnel records may not exist or may be in significant disarray, making this process of determining who is doing what job and where, and how many people are performing each role, a significant challenge.[13] In Libya during the Gaddafi regime, for example, records for how many people were in the security forces were deliberately kept from being accurately compiled. In the postrevolutionary years, a combination of ghost payrolls (for security personnel who had died or who had entered de facto retirement) and a botched attempt to integrate nonstate armed actors into the official security sector has left the country with no accurate record of how many people are employed by the defense sector, where they are located, and how many are actually working, not to mention what their training or experience is. Where systems for tracking personnel are absent or weak, the resources of the wider defense sector can be diminished by issues such as bloated militaries, ghost soldiers on payrolls, and defense planning impaired by the inability to gauge how many personnel are available and with what skills and training for any given mission.

For defense sector reform programmers, helping a country to obtain correct and sufficient data about existing defense sector personnel will likely be a top priority when building human resource planning capacity. One entry point for this may be the establishment of a centralized and comprehensive personnel management database. Establishing even a rudimentary record of defense personnel will allow for better resource decisions to be made at the strategic level and for planning to be initiated for (1) placing and keeping the right people in (or removing the wrong people from) the jobs the defense sector needs performed to achieve its goals; and (2) providing training and education to existing forces and new recruits to acquire, improve, and retain critical skills.

Second, the defense sector should have clear processes and policies for personnel management. Clear and formal personnel management processes and policies are necessary for the defense sector to efficiently manage all the people determined necessary to achieve the country's defense goals. Effective organizational management depends largely on members of the ministry of defense and the general staff, who can provide leadership for the professionalization of defense sector personnel. Personnel management must include mechanisms to address important challenges such as continuity of knowledge in the defense sector and effective communication with employees both vertically and horizontally. At a minimum, effective defense personnel management structures will include (1) merit-based *recruitment*; (2) transparent *compensation*; (3) performance-based *retention*; and (4) standardized, mandatory *retirement*.

The defense sector will function best if *recruitment* processes generate the right candidates for the job. With limited budgets in many of the states where defense sector reform will be a priority, hiring the right people for the right jobs—and, indeed, removing unqualified people from jobs they currently hold—will be a first, critical element to yielding return on investments in defense personnel.

Recruitment must be based on uniformly applied standards, with criteria reflecting the necessary skills to perform the given role. A prerequisite to setting recruitment criteria is clearly defining the roles and responsibilities for each position. In democracies, such employment standards and criteria are usually publicly available. Clearly defined, uniformly applied, and transparent recruitment standards can reduce the likelihood of hiring decisions based on corruption or nepotism that may result in unqualified individuals taking on roles they do not have the skills for and cannot perform adequately, or certain societal groups, such as minorities or women, being put at a hiring disadvantage.

To ensure consistency and transparency, the defense sector must have processes and policies that codify the standards of merit- and quality-based recruitment and advancement. For the armed forces, recruitment policies may include standards of physiological, medical, and/or psychological preconditions to ensure a person can carry out missions in the field. In the case of commissioned officers, many democratic armed forces require a higher education degree from a civilian university or military academy as a prerequisite. The recruitment of civilian personnel may also require a certain level of education, a specific degree, specialized skills, or a certain number of years of experience in a similar role.

Countries must also have a system for vetting incoming potential hires and military recruits for past disqualifying events, such as criminal convictions or disorderly conduct.[14] For postconflict countries undergoing defense sector reform, the vetting of recruits for involvement in past human rights violations may be necessary. Where recruits from former nonstate armed groups are being integrated into the armed forces of a postconflict country, specialized training must be designed to help recruits develop new skills and inculcate loyalty to the country's defense sector.[15]

A challenge to recruitment in many countries where defense sector reform will be a priority is the relatively small pool of skilled workers who already have the necessary skills or qualities. To this end, in addition to defining standards for recruitment, programming should aim to establish education and training systems for new recruits and existing personnel that address these skills gaps. Programs should also keep in mind the limitations of the availability and capability of the existing human resources of the defense sector; overly complex programming is likely to fail if the right people for the personnel management jobs do not yet have the necessary skill sets.[16]

To attract the best people, the defense sector must be able to provide competitive and stable *compensation*. Compensation for civilian defense and military service personnel should be standardized according to clearly defined and transparent criteria. Depending on the country, compensation may include salaries, benefits for service members and their families, financial compensation after retirement, and/or provision of access to veteran healthcare after separation. This requires that the defense establishment project its budgetary needs to compensate its people, effectively communicate with and obtain those resources from the government, conduct financial planning to budget for and distribute compensation, and implement a system to consistently pay personnel.

Many of the countries for which defense sector reform is likely to be a priority lack sufficient funds for the defense sector, limiting their ability to provide adequate or competitive compensation or to base, house, feed, and train military personnel. The lack of funds can also leave the defense sector unable to pay its workforce consistently. Where the defense sector is unable to pay its people, low morale or weak loyalty may present a security threat. Poor or unreliable pay may also tempt personnel to supplement their income through corruption.

To *retain* the best people and ensure that middle management and leadership positions are filled by those qualified and able to do the job, promotion policies and processes should be based on clearly defined criteria and be uniformly applied to all personnel who qualify. For example, according to one defense scholar, general principles of military promotion should be

> based on the hierarchy of assignments and military ranks structured in accordance with law . . . [and] regular (annual) evaluation, on preparation for the next assignment, and on graduation from a certain school or course; [o]ne assignment level [should match] only one military rank and the promotion can be—in peacetime—only gradual, normally between a minimum and maximum waiting period of time related to the assignments; [h]igher command positions should be accessible only through a sophisticated system of international, staff, and other professional assignments; and . . . promotion of officers should be coupled with a territorial mobility while in the case of other ranks this request is connected only to warrant officers.[17]

Systems for regularly evaluating performance should be based on uniformly applied, clearly defined, objective criteria. In the armed forces, leadership positions will be filled internally, and therefore tracks for promotion should include developing the skills that lower-level personnel will need to advance into those positions: "Unlike other [organizations], the armed forces do not import senior people from outside their own structures. The leadership (officers and senior NCOs) have to be provided from within. It will be necessary, therefore, to have procedures which can identify personnel for advancement and provide them with the necessary training to enable them to fill more senior positions."[18] Merit-based promotion and recognition are important for demonstrating the rewards of and respect for good work, which is critical both for building loyalty and for setting standards for civilian and military personnel.[19]

Evaluations are at the heart of the promotion and qualification system, but they also include the basis for decisions about unacceptable behavior or actions and consequential demotion or dismissal. Reprimanding of offences (both military and criminal) in the defense sector must be regulated by established codes of conduct, codified into law, understood by those to whom they apply and those who will use them to regulate, and adhered to uniformly. Only when there are consequences for crimes that benefit the individual, such as corruption, will there be a strong reason for defense personnel to actively change and avoid those behaviors.[20] To this end, laws must establish the democratic rights of military personnel as civilians in uniform and with specific attention to how soldiers are disciplined.

Finally, the defense sector must have standard processes in place that designate when and under what specific circumstances defense personnel and members of the armed forces can and must retire. *Retirement* is important and necessary for the defense sector to be effective. In the case of the armed forces, for example, retirement ensures that the active military is composed of young and able people, as well as allowing for career advancement through the constant rotation of retiring personnel out of the top positions so that low-level and mid-level personnel can advance their careers, making their retention more likely.[21]

In some countries where defense sector reform is necessary, bloated militaries are large in personnel numbers but ineffective in practice. In such cases, the military tends to supplement the salaries of a significant swath of the population, leaving the defense sector without the resources or the open positions to hire appropriately qualified people. Individuals are often allowed to stay in their positions and keep their salaries long after they can no longer effectively do the job they are paid for. Following the fall of the Soviet Union, many Eastern European countries inherited a legacy problem of bloated militaries. Large numbers of defense personnel had to be initially cut and retirement processes and policies had to be created and instituted to ensure the bloating did not reoccur.[22]

Processes for retirement may include setting policies for the number of years a recruit must serve in the military before he or she can retire and specifying circumstances for early retirement or parameters for retirement due to disability. Retirement processes allow for the exit of personnel from the military to be anticipated and managed and provide economic security for defense personnel, particularly those who serve in the armed forces, once they leave the service. Further, the benefits inherent in a country's retirement policies can serve as a critical element of a country's recruitment strategy. Such processes will require financial planning to account for the funds to cover benefits such as pensions or, in the case of military personnel wounded in duty, disability payments and veteran healthcare services. Such retirement benefits are important for the military, not least because hiring the best people for the job will require the defense sector to be a viable and appealing alternative to the private sector.[23]

Third, the defense sector must provide for the development of personnel skills through an established system for education, training, and exercises. While hiring the right people for the defense sector depends on recruiting for existing capacity, it is the responsibility of the defense sector to provide continuous development opportunities for each person in the armed and civilian defense workforce to improve existing skills, adapt skill sets to meet changing environments, and gain new skills for career advancement.

Developing personnel skills requires formal systems for training and education; the physical spaces to conduct training and education (such as academies); people with the appropriate pedagogical skills to deliver the training courses; the ability to track what education an individual receives in his or her career path; and development of appropriate, effective curricula. For new recruits in the armed forces, for example, this development starts with orientation, education, and training to provide for the basic skill sets—cognitive and physical—necessary to perform the job at hand. It includes the provision of training and education throughout an individual's military career to ensure the development of the skills or knowledge necessary for an individual to move vertically or horizontally within the military structure.

At the officer level in the armed forces, both commissioned officers and NCOs hold positions of authority in the command-and-control structure, making adequate management techniques and leadership training an important element of the officer-level curriculum. Poor people-management skills can hinder the effectiveness of the defense sector: "Security and justice institutions often . . . make poor use of the human resources they have. Overhauling the selection, appraisal and supervision of officers and officials can be central to effective and sustainable reform [programs]. Poor personnel management can result in low morale among officers and mean that [defense] services do not make the best use of their existing staff."[24]

Equally, officers and senior commanders are frequently expected to develop strategies and make strategic decisions, tasks that require a cognitive critical and creative thinking skill set. Developing training and education for critical thinking is particularly difficult in countries where systems have actively discouraged individual decision-making.[25] The OECD DAC Handbook also notes that middle management should not be overlooked for senior or new recruit training programs; middle managers are often key drivers of change processes. "Middle managers need the capacity to manage processes of change and steer through reforms." (Additional information on developing strategic thinking is presented in *Goal 9: Strategy Generation*).

In many of the postconflict countries where defense sector reform will be implemented, training systems will have to be rebuilt or built from scratch. In Libya, for example, systematized training for defense and security personnel completely collapsed in the post-Gaddafi era. What training now exists for new recruits is largely on-the-job learning; some officials (and usually the wrong ones) also attend internationally provided training sessions, but these can be based on what the donor has the funding for, rather than on what the Libyan forces need, and an individual session can have little or no relationship to preceding and following sessions. The result is an ineffective defense sector, composed of unskilled and untrained personnel, with little to no real authority or power on the ground.

For defense sector reform implementers, helping the host country to establish its own system to identify what training is necessary, plan for how and when individuals will receive that training throughout their careers, and establish standardized training and education systems across the defense sector is a significant challenge. For many implementers, though, delivering training and education programs is the area in which they have the most experience and expertise to draw from when helping the host country to build capacity. One entry point that could help initiate this process is to assist the country in determining (1) what type of skills defense personnel need to perform their job; and (2) what type of training, if any, currently exists. From there, implementers can help the host country to start identifying gaps between what training is available and what is needed, and then determine priorities for curriculum development and training delivery.

In many defense sector reform environments, these three principles—strategically planning for, managing, and developing the right people in the defense sector—will be hampered by significant systemic challenges. For example, many new or transitioning democracies may also have a legacy of personnel management mechanisms tailored to keep the armed forces serving the interests of the former regime while suppressing threats (e.g., coups) to authoritarian leadership. In such cases, there is rarely a culture that promotes individual career development for defense employees or members of the armed forces or that is concerned with enhancing the effectiveness of the force to better protect the well-being and rights of the civilian population. Bringing the entire defense personnel management system in line with the necessary elements of a democratic defense sector often involves actively unlearning the practices of an authoritarian past, both within the defense establishment and in accordance with changes in the broader sociopolitical environment. In postconflict countries, the challenges may be even more pronounced. Establishing a system to hire and retain the right people for the defense sector may well require building an entire personnel management system from scratch and doing so within the confines of extremely limited availability of skills or resources.

In other countries, a deeply embedded system may already exist, but that system may be informal, based on personal relationships and unwritten normative rules, and one in which recruitment, hiring, and promotion decisions are made based on personal connections or favors rather than merit or skill. In Egypt, for example, the military is the main arm of the government and owns all of Egypt's defense industries. Many people want to join the military to have access to those industries. Another example is Libya, where many citizens who took up arms to fight during the 2011 revolution were subsequently given positions in the defense forces and the defense ministry as a reward for aiding the overthrow of the Gaddafi regime. Others were given jobs in the defense sector or armed forces as part of a failed effort to reintegrate the many militias that the revolution spawned, regardless of their military training or experience. This, coupled with a crippled training system, has resulted in a defense sector consisting of individuals who do not have the capacity to carry out important security functions, such as guarding the country's borders. Ineffective personnel management is at the core of why the GNA cannot effectively manage Libya's major defense threats. The result in both Egypt and Libya is a civilian defense and military workforce and leadership core made up of people that do not have the education, skill sets, experience, or sometimes even the physicality to effectively do the jobs for which they were hired. That said, the systems are usually deeply rooted and their practices often culturally acceptable. The kind of institutional cultural change that is needed could take decades, if not generations, to achieve, so the challenge for program designers will be to identify strategic entry points to at least initiate the longer-term change process.

Applying the Goal 8 Principles in Practice

These three principles demonstrate what defense sector reform practitioners can expect to see across a broad variety of democratic defense sectors if the defense sector has the appropriate policies, rules, and practices to plan for, identify, recruit, develop, retain, retire, and manage civilian and military personnel. In practice, however, the application of these principles will vary considerably depending on the context of the country undergoing reform. The following examples from various reform environments demonstrate how the application of personnel development and management reform interventions may vary, with the aim of helping defense sector reform practitioners think through how to translate the principles into possible activities, depending on the status of the workforce and its supporting institutions.

1. The defense sector plans for an effective and appropriate workforce to staff the military and its supporting functions.

The effectiveness of any defense sector relies heavily on purposefully determining what personnel it needs (in terms of numbers, composition, and skill sets) and then getting the right people in the right positions with the right capabilities to achieve the defense goals of the state. This is the *planning* function of personnel management, and it is critical for the development of an effective and appropriate workforce to staff the military and its supporting functions. Planning will require the defense sector to have the capacity to analyze current and potential future operating environments to inform decisions about defense personnel composition and then to develop road maps to recruit, hire, and retain to achieve that ideal composition. (This subject is further discussed in this guide under *Goal 6: Defense Planning.*) In countries where defense sector reform will be a priority, however, the capacity to plan may be limited by, for instance, ongoing crises or a lack of time, personnel, and other resources.

In countries recently or still mired in conflict, defense resources are likely to have been devoted exclusively to the ongoing conflict, with long-term planning for anything, including personnel, being entirely neglected. Further, internal conflict often leaves countries with personnel issues related to deep-seated regional or ethnic clashes of those previously in power or those who have won power through conflict. In Libya, for example, following the revolution in 2011, the defense sector's workforce largely broke down due to the presence of both a distrusted "old guard" and the revolutionaries who largely had no defense experience but who had fought during the revolution in militia groups. Those who had been part of the Gaddafi defense sector were pushed out, but the revolutionaries, who were trying to build a new type of state, lacked the experience, skills, and education to undertake the roles they assigned themselves in the new defense sector. As a result, Libya's current defense workforce is not the result of purposeful planning, and most of the personnel lack the skills necessary to maintain a military able to defend the country.

Compounding this issue is the fact that planners need to know—even if only approximately—how many people are currently employed by the defense sector, where they are located, what kind of training they have received, and so forth. In Libya, however, the integration of nonstate militia members into the state's formal security forces has left the defense sector with no record of how many people are on its payroll or how many of the people carrying out formal security roles are trained or educated in skills needed for the role they are performing. For example, prior to the revolution, Libya's General Administration for Coastal Security (GACS) had approximately four hundred trained individuals in the force. Following the revolution, however, over two thousand untrained individuals were added to the GACS to integrate militia members into the official security forces.[26] These individuals had no training or background in defense, and many of them have never shown up for duty, even though they collect a paycheck from the government. With the defense budget being depleted by payments to militia members as "defense personnel," funding to establish a formal system for personnel management planning does not exist.

In cases such as Libya, where no defense personnel planning has recently taken place, implementers will likely have the difficult task of establishing a personnel planning system from scratch. To do so, it will be important to first establish a baseline of existing personnel. One entry point is to conduct an assessment of the host nation's existing defense personnel to establish, for example, how many people are on the defense sector's payroll; how many of the people on the payroll are actually actively working (and, in some cases, are still alive); what the payments are and what the amounts are based on; what the people being paid are being paid to do; who is distributing the payments and how; how many of the people on the payroll have received training and what that training entailed; and so on. Only with this information in hand can planners decide whether, for example, the force needs to be downsized, and then develop appropriate plans.

An example of this process of taking stock of current capacity took place as part of the defense sector reform efforts in Georgia. In 2007, the Georgian Ministry of Defense abolished the personnel system's Soviet "specialty codes" for job descriptions and instead developed and implemented a personnel management system that has job descriptions and requirements for each position in the Ministry of Defense and the Georgian Armed Forces. As part of the process of switching all existing personnel to the new system, Georgia's J1 (organizational planning and personnel management section) created and led commissions that required each active service man and woman to attend an interview, provide proof of their education or specialties, have their personnel files reviewed, and then be granted a job code based on their knowledge and experience. The new personnel information files were cataloged for further data and career management.

Georgia is an example of a county that has received decades of international assistance to reform its defense sector. In such relatively established democracies, planning for defense sector personnel and force structure is far more likely to be an established part of the broader defense planning system. In Georgia, there is a chapter in the Law of Georgia on the Defense of Georgia dedicated specifically to the "types and composition" of the armed forces that delineates the types of services (land forces, air forces, national guard, and special operations forces) and their composition.[27] Force management, including force design, deployment, and trainings, is based on the Strategic Defense Review and Implementation Plan, which was developed in accordance with the Georgian National Military Strategy and Threat Assessments and aims to ensure that the current force design, deployment, and trainings are adequate to address the external and internal threats and defense objectives of the state.[28] The Strategic Defense Review provides four-year force-planning cycle and manning priorities to ensure, as one general staff official put it, that the defense sector knows "how many people and what skills, and where we need [them] for the next four years. So, our recruitment, assignment and distribution, professional development and promotion policy and plans will be linked to this four-year force planning cycle."[29]

In cases where the defense planning system is well established, implementers are likely to focus programming on specific parts within the system that may need to be updated or improved. For example, despite its robust system for planning for the right people—including such programs as the Georgia Defense Readiness Program, which aims to train Georgian troops to meet defense requirements—Georgia's defense personnel planning capabilities are limited by budget constraints.[30] Programming could be designed to identify efficiencies given those constraints; alternatively, budgeting challenges (which are addressed separately under *Goal 7: Financial Management*) could be addressed by adopting a better planning system. If planning is limited by a lack of processes or by skills gaps in personnel who are assigned to manage this task on behalf of a recipient defense sector, other entry points could focus on process streaming the existing planning process and then developing new systems or procedures to address any identified gaps. If skills gaps are driving inefficiencies, specialized training on human resource planning may be a place to start.

2. The defense sector has clear processes and policies for personnel management.

One of the most important aspects of having the right people is having an effective defense personnel management system. The defense personnel management system must have clear and formal processes and policies that ensure merit-based recruitment; transparent compensation; performance-based retention; and standardized, mandatory retirement. In countries where defense sector reform will likely be needed, the personnel management systems are likely to exist in some format but will often need significant assistance to help them establish or reform their existing recruitment, compensation, retention, and retirement systems.

In postconflict countries, personnel management systems will likely be outdated or missing. In Libya, for example, processes and policies for personnel management are informal and undocumented. This is a legacy issue tied to how personnel were managed during the former authoritarian regime. First, all formal recruitment, performance evaluation, and promotion processes were frozen by Gaddafi in 1986. So, for decades before the revolution, these processes were informal, ad hoc, and chaotic. Second, Gaddafi designed the state to prevent coups and to ensure the maintenance of his personal control. What official procedures, rules, or processes do exist on paper are outdated and are not followed or implemented in practice. Compounding this issue for the defense sector is that, due to an inability to provide formal security, the Government of National Accord provides many of the nonstate militias in Libya with salaries to conduct certain security functions without oversight, which has further institutionalized the informality of personnel management and diminished the value of any formal policies, which could not be enforced.[31] The result is that recruitment is largely based on knowing someone on the inside, compensation is not consistent, retention is dismally low, and retirement is not enforced because payments are made regardless of whether individuals show up to their job.

In environments where defense personnel management systems either do not exist or exist on paper but not in practice, practitioners may need to develop programming to help build a personnel management system largely from the ground up. It will be important for such activities to be tailored to the technical level of the country at hand and not to expect (incorrectly) that the importation of technology used by Western nations to manage their personnel will be an easy fix. For example, many Western personnel management systems rely heavily on human resources software and computer-based data storage. In countries on the lower end of the defense sector capacity spectrum, however, the infrastructure to support such systems may be missing: if there are no computers, no power, or no internet, these advanced systems will not work and cannot be sustained. Instead, programmers must determine what personnel management systems (if any) the country has in place already and design programming to improve the existing capacity in ways that can be both implemented and sustained.

If, for instance, the country has a paper-based personnel records system, but the records are stored in disarray and are not up to date, one of the first tasks will be to help the country to assess what records it has, which records are missing (which may be all of them), how to collect missing records or personnel information, and how to organize records so that current personnel files can be easily and systematically identified when necessary. Such a small step toward improving organization will be critical for determining how many people are on the payroll, who should be removed, and what recruitment is necessary to fill gaps in the current force structure.

Even in countries that have received significant foreign assistance, personnel management is frequently a low priority, despite having a significant impact on the country's ability to provide defense. For instance, the Iraqi Security Forces, which has received years of foreign assistance, continues to struggle with how it onboards, trains, supports, pays, keeps healthy, and retires its armed forces. Defense jobs are rarely filled based on an individual's skill set matching the requirements of a particular role. Rather, family, tribe, political party, and sectarian affiliations take precedence for hiring and promotion, often in that order: "hiring within Iraq is based primarily on connections and networks; the priority for filling an open position is often to gain security for a family member or friend."[32] The result is a defense sector composed of individuals who are unqualified to do their jobs or uninterested in doing their jobs well. Compounding this issue, widespread corruption often prevents the right people from getting the right job. Misconduct at all levels of the defense sector is regularly reported in the media, including misconduct related to hiring practices.[33] For example, the United Nations reported that "slightly more than a third of all civil servants in Iraq learned that a particular job was available through friends and relatives (21.9 percent through friends and relatives working in the same ministry and a further 12.3 percent through friends and relatives working elsewhere), while about a quarter learned of a job opportunity through the centralized appointment system for graduates and 13.7 percent through public advertisements."[34]

In Mali, as in Iraq, military recruitment and personnel planning remain a challenge despite significant foreign assistance. Defense personnel tend to be chosen based on patronage instead of technical qualifications. The lack of qualified personnel has led to a force made up of individuals who do not have the skills to fulfill the roles that the defense sector needs them to perform. This reportedly had a significant impact on Mali's operational readiness ahead of the events that led to the north of the country being overrun by rebels in 2012.[35]

In countries where the defense recruitment process is rife with corruption and nepotism, changing the system will be difficult and controversial. The system in place is likely to benefit many within it, including, often, those at the highest levels. For implementers, changing the recruitment culture will require generating buy-in among key stakeholders by highlighting the benefits of a competitive recruitment process, notably the increased efficacy of the force.

Other incentives, such as access to training or opportunities for promotion, may also be important for shifting that institutional culture over time. Designing a new recruitment process will likely require technical advice and assistance to review the existing process(es), identify gaps, and develop the appropriate processes and policies to address those gaps. Entry points will likely be determined by priority gaps and needs. If the defense sector struggles to retain personnel, initial programming may focus on developing more robust retention policies; if recruitment lags, expanding the method and means of recruitment may be a better place to start. Quick wins early in the process may also help shift stakeholders' support for the proposed changes, particularly if those changes are expected to be highly disruptive or disempower actors who benefited from the old system.

In other cases, particularly in more established democracies, the defense sector is likely to have a significantly more robust personnel management system. In Georgia, for example, the defense sector personnel management system—including recruitment, compensation, retention, and retirement—is well established under the Law of Georgia on the Defense of Georgia. Defense personnel policies are reflected in the Human Resource Management Concept (which provides an overarching concept for the main principles and directions of military and civilian personnel management in the Ministry of Defense and Georgian Armed Forces); the Military Personnel Management System Development Strategy (which details the military personnel management system); the Civilian Personnel Management System Development Strategy (which details the civilian personnel management system); and the Vision on Institutional Development of Human Resource Management in the Ministry of Defense (which assesses the status of personnel reform in 2017 and discusses the directions to enhance personnel management mechanisms).[36]

Despite its robust structure, Georgia's defense personnel management is not without problems, particularly with the recruitment and retention of a new professional force.[37] As part of their efforts to become NATO-compliant, the Georgian Armed Forces are aiming to transition from a conscription military to a professional army to improve the general skill level of the forces and bring the defense sector in line with other Western militaries. However, the transition of any country's army from a force made up of conscripts to a professional military will likely result in a reduced force size and a more expensive force.[38] The recruitment process includes using open competition for recruiting, announcing vacancies on the Georgian Civil Service Bureau–managed website, posting vacancy announcements that contain job descriptions with qualification requirements, allowing any adult citizens of Georgia who meet the job requirements to apply, and calling for interview within two weeks the applicants who meet the requirements for the interviews or testing. The Competition Commission—which is chaired by the deputy minister and staffed with the head or deputy of administration, the head or deputy of the human resources department, the head of the legal department, and the head or deputy of the department where the vacancy is—then interviews short-listed candidates and selects the most suitable candidate based on the scores the commission members give to the applicants. The name of the applicant with the highest score is submitted to the minister for appointment to the vacant position. This system is good, even by Western standards, but the problem in this case does not lie in the system itself. In Georgia, recruiting new members to join the professional force by choice is a significant challenge because, despite the Georgian military's relatively good standing in society, joining the Georgian Armed Forces is not considered an attractive career option by the young people that the force most needs.[39] Despite efforts to address the issue—for example, by making pay scales more standardized and fair—the inability to recruit enough of the right people has hindered the military's professionalization.

Beyond the armed forces, retention of civilian experts in the Georgian Ministry of Defense has also been challenging. The majority of senior staff positions in the ministry are political appointments and each time a new minister of defense is appointed (12 different ministers have held the position between 2004 and 2021), senior civilian positions are also changed.[40] Further, those who are appointed are often not security experts (e.g., one minister and her staff were former lawyers), which leads to a steep and time-consuming learning curve every time the new staff come in; frequently, by the time they are beginning to understand the role, they are replaced by a new cadre.[41] This has left the ministry without continuity of knowledge and without experienced senior staff. Although there are lower-level civil servants who do remain in their roles across administrations and could be a good source for knowledge continuity, these roles are beholden to the senior appointees and there is little job security as a result (new administrations often replace staff as they see fit), making retention at the lower levels problematic, too.[42]

In cases such as Georgia, where the personnel management system is fully established and relatively functional, programming will likely be focused on specific areas where functionality is missing or could be improved. An initial assessment that involves interviewing staff at all levels throughout the ministry and armed forces is a good entry point for donors to determine where existing staff see gaps or problems in the day-to-day activities of the existing personnel management system. Based on this assessment, activities can be identified to target specific problems in retirement, compensation, recruitment, and retention practices and policies. Colombia, for example, has an established defense personnel management system, within which the recruitment and hiring of personnel is conducted by the Department for the Administration of Public Service following the regulations established in the Civil Service Law. In Colombia's case, although guidelines and protocols for entering the civil services are clear, there is widespread failure to abide strictly by the regulations.[43] This is an example of a specific area where defense sector reform efforts could target a part of the existing personnel management system to strengthen its functionality.

3. The defense sector provides for the development of personnel skills through an established system for education, training, and exercises.

Hiring the right people for the defense sector depends on recruiting for existing capacity, but the continued readiness of the force to respond to current and emerging threats requires continuous development opportunities for defense personnel to improve their existing skills, adapt skill sets to meet changing environments, and gain new skills for career advancement. Developing personnel skills is achieved through formal systems for training and education; physical spaces to conduct training and education; people with the appropriate pedagogical skills to deliver the courses; and the development of appropriate, effective curricula. In the types of countries where defense sector reform is likely to be necessary, defense training and education will vary widely from complex established systems, on one end of the spectrum, to ad hoc training, on the other end. In some instances, training programs exist but are no longer being delivered due to a complex array of circumstances, including ongoing conflict or lack of resources. In those instances, personnel frequently are forced to *learn on the job*.

Countries that have recently been mired in conflict or that are still engaged in conflict are likely to have threadbare or no formalized and functional training and education systems. In the case of Libya, for example, since the revolution in 2011, formal training for the defense and security forces associated with the UN-backed Government of National Accord is ad hoc at best and nonexistent in most cases. Under the Gaddafi regime, the Libyan defense establishment had three levels of training and education that fell under the Military Training Directorate: first, the Military Training Centers; second, the Military University; and third, the Military Academy.[44] These three establishments had within them multiple disciplines, several colleges, and constituted a relatively high-level training and education system. After the revolution in 2011, however, Libya's military training system largely broke down. Today, a few parts remain functional in a limited capacity, with some facilities open only for administrative staff and others delivering ad hoc training. The academies, training centers, and universities that provided training under the Gaddafi regime are largely nonoperational due to a lack of funding and equipment, administrative grid-lock, and security threats. The Government of National Accord's Ministry of Defense has within it a department of planning that is meant to be responsible for curriculum development, but new recruits to Libya's defense services—most of whom have no military or security background—are generally expected to learn on the job, rather than receiving even basic training in specific functions, tasks, and procedures.[45]

In postconflict environments where the training and education system has largely or completely broken down, programmers will likely be faced with helping the country to develop a defense training and education system from scratch. This will include determining if any of the remaining elements of the old system are worth salvaging and will likely require conducting an assessment to gain a clear picture of the defense landscape, as the chaos of conflict is likely to have left very few people with an understanding of where the system stands. Once there is a baseline understanding of what exists, one entry point for pro-grammers could be to help the host country develop a blueprint that details what the larger defense training and education system should entail, how it will function, and under which authorities. Having a blueprint of the broader defense education and training system will allow the creation of such a system to be broken down into sequential activities that gradually build out the system.

On the other end of the spectrum are relatively advanced democracies such as Colombia and Georgia. In such environments, implementers are likely to find robust training and education systems. Colombia, for instance, has three main defense training schools: a military school for cadets, a center for military education, and a superior war college. Professional soldiers obtain technical training; noncommissioned officers are granted a technological degree; and of-ficers receive a professional degree. Importantly, all members must first pass specific requirements for entry into the force.

Georgia also has a robust defense training and education system. The Law of Georgia on the Defense of Georgia states that the Ministry of Defense "shall be responsible for training and developing Armed Forces."[46] Within the Ministry of Defense, the Training and Military Education Command holds the respon-sibility to "to provide the personnel of the Defence Forces of Georgia with mil-itary professional education, training exercises and assessment of sub-divisions, as well as to provide the Defense Forces of Georgia with the military doc-trines."[47] Several institutions under this command provide opportunities to members of the military to develop their skills and knowledge throughout their career. The schools and colleges provide "full-spectrum training of the spe-cialists (private, corporal, sergeant, officer) as well as basic combat, mountain,

sergeant and pre-deployment training/retraining."[48] One training and education official noted that "the objective is vertical and horizontal synchronization
of all military educational courses, and full compliance with . . . military policy,
strategy, and priorities."[49]

Each course within the defense education system has a course assessment
system, and the professional military education system includes officer career
courses (e.g., the junior officer course, captain course, command and staff
course); a joint training and education center; a common combat center; and
a doctrines center, which integrates all doctrinal developments into the military education curriculum. Ultimately, Georgia's military education system
focuses on developing leadership and decision-making skills to prepare officers
and NCOs for decentralized management and to hone their ability to make
decisions in complicated or stressful situations.[50] Importantly, all ranks in the
defense forces must complete predetermined levels of training to qualify for
advancement and promotion.[51] Professional development paths are elaborated
for each military branch, and defense personnel who want to be considered
for promotion must meet specified criteria for lengths of service, staff or field
experience, and career education.

In cases where the military education and training system is institutionalized and relatively robust, as in Colombia and Georgia, programmers will likely
be tasked with addressing a specific skill or knowledge need or gap. Alternatively, if the training needs are broader in scope (e.g., there is a need to strengthen
training for new recruits), programmers can conduct an initial assessment of the
existing curriculum and pedagogy to determine where there are weaknesses. For
example, some countries may have a potentially effective system in place, but
the courses and course materials may be outdated, focus excessively on theory,
or rely too heavily on rote learning. If systems are overly reliant on rotating
instructors who are not prepared to teach or train, defense sector training institutions may benefit from the introduction of an instructor certification course
or program to improve the quality of training delivery.

CHAPTER ELEVEN

GOAL 9
Strategy Generation

The ninth goal of defense sector reform is the effective *generation of strategy* for the defense sector. Such strategy should, at a minimum, link a country's defense objectives to the military and other resources at its disposal to achieve those objectives. Put simply, a strategy defines the *what* and *why* of the national, defense, or military objectives of the country, as well as describing in broad terms *how* a country will employ its military and other resources to achieve those objectives.

Most nations, small and large, have a strategy of some description, at least on paper. Those strategies will vary significantly from country to country. Depending on the complexity of a country's defense forces and coordinating institutions, some countries will have only a single overarching strategy for the entire security sector, whereas others will have not only a security strategy but also a defense strategy, a military strategy, and many other supplemental, subordinate strategies that tie into the high-level strategies. For example, in addition to its national security strategy, the United States has a combatant command strategy for each of the geographic and functional military commands, a ministerial-level strategy for the Department of Defense, and a service-level strategy for each of the armed forces (Army, Air Force, Coast Guard, Marine Corps, Navy, and Space Force).

In a hierarchical sense, a country's *national security strategy* is the overarching strategy from which all other security-related strategies, including the defense and military strategies, are derived. It defines security objectives for the entire country, in line with the nation's long-term core interests, values, and beliefs, and includes all the instruments of national power, not just the armed forces. This top-level guidance broadly articulates how security institutions and forces will advance the country's overarching strategic vision.

Subordinate to the national security strategy, a country's defense sector may also have a *national defense strategy*, which provides guidance on force employment, force planning, force design, posture, programming, and other activities, in line with defense priorities that complement the national security goals. A

137

national defense strategy serves to frame guidance for the defense sector and structures prioritization of the use of defense resources and the planning of activities for defense forces to give a country the military advantage, to build stable foreign relations (e.g., to affirm alliances or demonstrate intent of good neighborly relations), and to deter potential adversaries. Strategies subordinate to the national defense strategy tend to be more operational in nature; for example, military strategies are focused on how the armed forces can contribute to specific operations in support of the defense strategy.

The importance of strategy is emphasized to varying degrees across the defense sector reform literature. The U.S. Department of Defense's DIB Directive lists "strategy" among the principal functions and duties of effective defense institutions, and the UN Defence Sector Reform Policy notes that the host nation's defense sector reform plan should be "derived from a recent national security policy/strategy," adding that should a security strategy not exist or be outdated, implementers should assist the national authorities in developing one.[1] While having a defense or security strategy documented is important, without building the capacity of a country to generate strategy, such documents will rapidly become outdated and ineffective. Defense sector *reform*, as opposed to defense sector *assistance*, will require institutionalizing the recipient country's ability to develop and continuously carry out a broader *strategy generation process*.

The OECD DAC Handbook identifies five steps for developing a national security strategy: (1) conducting strategic environmental analysis and identifying a national vision; (2) analyzing and prioritizing current and future threats to achieving that national vision; (3) determining and prioritizing national capabilities required to successfully address key threats and deliver key services, including defining acceptable risks and considering affordability; (4) determining lead responsibility for delivering each capability and establishing means of ensuring effective coordination, accountability, and oversight; and (5) establishing a national security strategy.[2] The UN Defence Sector Reform Policy does not detail what the strategy development process entails, but it does describe a "national policy process," which includes several important elements of a strategy generation cycle, including (1) "appraisal and review of the national situation in political, economic, social and military terms and in close relation to the regional and international contexts;" (2) "identification of national objectives based on national interests and aspirations"; and (3) a government "directive that will encompass all the ministries and will allow their planning for the national security policy and/or strategy."[3]

The NATO PAP-DIB alludes more specifically to the importance of the broader strategy generation process, calling for the development of "effective and transparent arrangements and procedures to assess security risks and national [defense] requirements," which are important aspects of strategy generation, as well as the development and maintenance of "affordable and interoperable capabilities corresponding to these requirements."[4] Yet, the literature largely falls short of describing what institutionalized processes are required at each phase of strategy generation—for example, development and intelligence analysis, projection, articulation, vetting, communication, implementation, adherence, revision, and adaptation—for the defense strategy to be effective.

Across the defense sector reform literature, there is also very little guidance on developing the skills, particularly the cognitive skills, necessary to carry out the various steps of strategy generation. Although most documents acknowledge that a host country–led process is critical to the institutionalization of any strategy, the host country personnel's *capacity to think strategically* is often assumed to exist or is overlooked. In many of the countries where defense sector reform will be a priority, individuals may have been discouraged from conducting analyses, expressing opinions, and making decisions and are unlikely to have received training in decision-making. The NATO PAP-DIB curriculum guidance does note that a recipient country's defense education system should add as a learning objective the ability to "[synthesize] the fundamental construct of the strategic thought process—the calculated relationship of ends, ways and means."[5] The curriculum guidance also lists understanding "the significance and components of National Security and Military Strategies," as well as developing "the ability to craft strategy . . . at all levels of the government" as key learning objectives for host country defense education institutions. Although the importance of having the ability to develop strategy is emphasized, the OECD DAC Handbook does not go into greater detail about how implementers can build the cognitive skill sets of individuals across the defense sector to be able to turn information into intelligence through analysis, to use that analysis to forecast the future defense operating environment and project what the defense sector may need to adapt to that environment, or to adapt a current strategy in real time to meet an unexpected crisis or change.

Guiding Principles for Design and Implementation of Goal 9

For strategy generation to be effective, countries must have a strategy generation process that ensures the strategy remains in step with developments in the operating environment, as well as people who have the strategic thinking skills to provide intelligence, conduct the analysis, project requirements for future contingencies, and so on. To this end, there are two principles that inform how strategy generation is achieved in democratic defense sectors.

First, strategy generation requires strategic thinkers who can appraise risk, calculate probability and consequences, mitigate danger, address challenges, and prepare for future threats. The development of strategy takes a certain set of cognitive skills that include the capacity not only to assess and analyze but also to project, predict, and offer creative solutions. Effective strategy generation therefore depends first and foremost on human capacity. To provide the defense sector with strategies to mitigate risks, address challenges, and prepare for future threats, defense ministry officials and officers in the armed forces need to have the capacity to appraise risk, calculate probability and consequences, and create strategy for their subordinates. In many cases, especially for military officers, this ability to strategize is often bound by tight time frames that necessitate quick decisions. Whereas doctrine serves to provide correct techniques and procedures for the use of equipment or behavior in the field, strategic thinking allows officers to make judgements that account for doctrine but are adaptive to circumstances at hand. Strategic thinking is also a critical component of responding effectively to those defense challenges for which contingency plans do not yet exist, such as the sudden rise of unexpected threats, unanticipated catastrophic events, and other crises that warrant rapid military response. Furthermore, strategy requires not only the ability to assess current threats but also the analytic skills to project what future threats may be.

This is no easy task, and it requires more than tactical and technical proficiency, although such proficiency may help tether strategic thinking to the reality of the likely battlefields for which the strategy is being generated. As U.S. Army researchers explain:

> Managing the complexity and uncertainty inherent in operational environments requires that military officers have the capability to anticipate and visualize possible future states; to consider the impact of actions on an array of geo-political factors; and to conceptualize what actions are necessary to shape future states in ways that benefit [national] interests. Moreover, the same skill sets are critically important within the [defense] institution itself . . . to determine priorities, set requirements, and allocate resources so the organization can adapt and transform to meet future needs.[6]

In most countries where defense sector reform will be a priority, there will likely be individuals who already have the skills required to develop strategy and project potential future requirements. However, these individuals may not have received the training necessary specifically to develop strategy, or they may work within a defense system that has no strategy generation process or that discourages strategic thinking skills. In some postauthoritarian countries, the kind of critical thinking skills necessary to create a strategy—conducting inquiries, demonstrating openness, expressing dissent, forming opinions—were likely discouraged (strategy generators can also be coup generators) or limited to a tightly controlled group of now deposed leaders. For that reason, in many of the countries where defense sector reform will be a priority, a culture that encourages strategic thinking will be a crucial but missing element. In other cases, select officers may have received their training or education abroad at a donor or allied country's defense education institution. In such cases, officers may have received education on how to think strategically and how historical military figures thought about strategy, but often this does not include training on the actual production of strategy.

Developing strategic thinkers within the defense sector will not be quickly or easily accomplished, particularly in countries where institutional cultures limited this type of thinking. Cultural change of that magnitude will take decades, if not generations, to effect. That said, there are certain types of programming that could initiate the development of a culture of strategic thinking within a country's defense sector.

To begin making strategic thinking routine and institutionalizing it as a process, programming can focus on building analytical and strategic thinking skills at all levels of the defense sector. A country's ability to institutionalize the capacity for defense strategy and sustain it should not be limited to a handful of exceptional officers. Programming can assist recipient defense sectors to incorporate strategic thinking elements into their education and training programs, particularly at midlevel and senior command schools. Curricula should emphasize questioning, reflection, discussion of ideas, and articulation of creative solutions, which in many rote-learning environments will entail a transformation of how military forces are trained and educated.

Second, strategy generation requires a formal system (or systems) for the development of that strategy. Defense strategy generation requires cyclical processes that together form a system. For that system to be effective, the defense sector must have the capacity to conduct each step or process at various levels within the defense sector and on a constant rotation. These processes include (1) the *formulation* of a strategy, based on realistically available resources and capacities; (2) the *articulation* of the strategy on paper; (3) the *approval* of the strategy in accordance with principles of democratic control; (4) the *communication* of the approved strategy to appropriate audiences, including users of that strategy; and (5) the *adaptation* of the strategy whenever necessary to meet evolving or sudden changes in underlying conditions or environmental circumstances.

Ultimately, the efficacy of the strategy will be determined by how it is implemented, including through the translation of the strategy's objective into plans and by the defense sector's adherence to the strategy.

Formulating defense strategy is tied directly to the first principle in that it relies on the human capacity to think strategically, assess, create, deliberate, and project or forecast. But there must also be a process for that development. Likely steps include (1) determining the delineation of roles and responsibilities between the duties of the civilian establishment and the military regarding strategy generation and approval; (2) assessing a country's security needs; (3) accounting for available resources; (4) determining a hierarchy of security objectives and priorities; (5) translating objectives into realistic plans for implementation; (6) projecting the country's future security needs and resource availability; and (7) developing any requisite subordinate strategies, plans, and corresponding military doctrine necessary to support the implementation of the overarching national defense strategy's objectives, dependent on the size and complexity of the force.

Articulating the defense strategy will likely come in the form of a written document. The articulation of the defense strategy is an important forcing function for guiding the defense sector in a unified direction. Without it, the various components that make up the defense sector, including each of the armed forces, would struggle to operate in unity and military effectiveness would be impossible. Although strategy documents vary significantly from country to country, an effective defense strategy document should include at least the following components:

- *A description of the guiding normative principles,* in line with the country's national interests, national aims, and national values.

- *A contextual overview of the current operating environment*—including enduring, critical, and potential challenges and threats—which explains and justifies the defense sector's strategic objectives.

- *An articulation of the objectives and subobjectives of the defense sector to counter threats and provide for the national defense.* These objectives must be based on realistic assessments of available resources and of what is operationally possible for the current operating environment, what challenges may arise in the near to midterm, and what innovations are required to respond effectively to future threats.

- *An implementation plan for the sector's financial, material, and human resources to achieve the objectives,* often including listing priority missions for the defense sector. The implementation plan broadly explains how the objectives will be realized. It is not, however, synonymous with defense planning, which is operationally focused and far more specific and is treated separately under *Goal 6: Defense Planning*.

- *Proposed reforms that the defense sector and its forces need to make to effectively implement the strategic objectives*, such as adopting new types of equipment, updating weapons systems, expanding training, and revising force structures.

- *A description of available resources.* Although many countries keep their defense budget classified, the defense strategy usually includes at least an acknowledgment of the available resources. Tethering available resources to the strategic objectives is a key step in making the strategy a practical rather than an aspirational document. The description of resources may, for example, account for how many forces are available, properly trained, and ready, as well as the number, condition, and availability of requisite equipment, weapons systems, and other resources.

- *A longer-term projection for future force requirements.* The length of time this covers will depend on the publication cycle of a country's defense strategy, which could be annual or every five years, but should not be so long that the strategy is outdated upon publication.

In addition to these components, strategies for more complex defense sectors may include elements such as alliance responsibilities and objectives, descriptions of the role of the country's armed forces, inclusion of the role of the country's defense industrial base, or nondefense sector elements such as the role of the country's diplomatic arm to prevent or reduce hostilities or the role of the police if an external threat has internal security implications, such as terrorism.[7]

Once it is articulated, the generation process must allow for *vetting and approval* of the defense strategy. In a democracy, the defense strategy must have at least input, and in certain cases also approval, from all the key stakeholders, which depending on the county could include the defense-related ministries, the defense forces, the executive branch (including the head of state), and the legislative branch (in the form of parliamentary committees of defense and, in their defense-related decisions, of security and intelligence).[8] The strategy may not specifically address the detailed functions of the defense sector but will serve as the starting point for all subordinate strategies relating to command and control, strategy and planning, defense intelligence, defense legislation, budget authorization, military justice, human resources, education and training, media and public relations, acquisition and procurement, provisioning, logistics, planning, deployment, and so on. The input and agreement of stakeholders are therefore essential to the pursuit of defense objectives and the effective implementation of the strategy.

The finalized defense strategy must then be *communicated* effectively and explained sufficiently to all stakeholders that are expected to follow its guidance. A defense strategy cannot be effective if those who are meant to follow it do not know that it exists or do not understand how it affects their day-to-day operations. Although the sensitive nature of defense strategies often necessitates classified sections, in a democracy it is important to also provide a publicly available summary.[9] In transitioning democracies, an ingrained culture of secrecy may make communication with other agencies and the public a challenging institutional change, but it is one that should be encouraged to ensure transparency and accountability.

Finally, the generation process must allow for effective *adaptation* of the strategy to ensure its relevance and applicability in evolving contexts. An effective strategy should be adaptable and flexible based on contextual changes and should be updated regularly to reflect evolving capacity realities. In practice, this means that a system must exist that allows the defense sector to assess, update, and innovate the defense strategy and subordinate strategies in line with new challenges, changing contexts, and potential future risks. In countries where limited capacity exists in terms of strategic thinking skills, adaptation will be a difficult process to institute, simply because of a lack of available manpower to conduct real-time analysis.

Once a defense strategy has been generated, the efficacy of the strategy will depend on what happens next; a strategy, no matter how well devised, is not useful sitting on a shelf. For example, the defense sector must be able to effectively operationalize the strategy by translating it to the program, project, and mission levels. The effectiveness of a defense strategy lies in its translation from words on paper to actions taken by the defense establishment—culminating in the realization of the defense sector's primary objective of keeping the population and its government safe from external, or where applicable internal, threats, as well as any other objectives stated in the strategy. This translation into practical application is largely the role of defense planning, which is discussed separately under *Goal 6: Defense Planning*. At the force level, implementation of the national strategy will include aligning force apportionment, assignment, allocation, and readiness to achieve the objectives of the national strategy. It may also include the development of campaigns regionally and globally to operationalize the actions of the armed forces in support of the strategic objectives. Translating the strategy through the defense sector down to the unit level will require a series of mechanisms in and of itself.

In countries where defense sector reform is likely to be a priority, existing defense strategies often tend toward unrealistic posturing with objectives based on aspirations rather than capacity and resources, while in other cases, documents that purport to be strategies are actually policy documents. Helping countries to implement a process for strategy generation will be possible only if the development of that strategy has been confined to the realm of the possible regarding available resources and capacities. For a strategy to be effectively implemented, it must be implementable. A strategy that accounts for priorities and is based on current and anticipated threats within the bounds of existing resources and capabilities (or accounts for the reasonable improvement of capabilities or increase in resources) is much more likely to be *adhered to* by those who are supposed to regard it as top-level guidance from which to derive all substrategies and plans. If the top-level guidance is unrealistic and not implementable, those carrying it out will disregard it out of necessity, and the actions of the various defense sector forces will be disjointed. The effectiveness of the defense sector will be compromised if each component is following a different vision for how best to defend the state.

These principles—having strategic thinkers and a strategy generation process—present two critical capacities for strategy generation that may require years, if not decades, of cultural change to institutionalize. For defense sector reform implementers, this necessitates finding entry points from which the first step in the long process of developing these capacities can be initiated. One such entry point is education. Where strategic thinking is not yet a norm, programmers should aim to design training or education programs that incorporate analysis, forecasting, prediction and trend analysis, and creative problem solving. Another entry point will be to identify the current strategy generation process (if one exists) and help design either new component processes or a strategy generation system to replace it. This design need not be complex, particularly if no such system exists. The recipient country must be able to execute and replicate the strategy generation process effectively within the limitations of available resources and manpower if it is to be sustained.

Applying the Goal 9 Principles in Practice

Strategy is a critical part of the defense process, and its effectiveness relies on having able strategists as well as a functional strategy generation system. While these two principles indicate how strategy can be effectively generated in general, their application in practice will vary depending on the country context. This section provides some examples of what strategists and strategy generation capacity may look like in varying environments to help practitioners translate the guiding principles into possible activities.

1. Strategic thinkers can appraise risk, calculate probability and consequences, mitigate danger, address challenges, and prepare for future threats.

Effective strategy generation depends on human capacity. Generating a defense strategy requires strategic thinkers—individuals who have the ability to appraise risk and calculate probability and consequences, while mitigating danger, addressing challenges, and preparing for future threats. The cognitive skills that will be required include the capacity not only to assess and analyze but also to project, predict, and offer creative solutions. In many countries where defense sector reform will be a priority, the extent of these skills among ministry officials and military officers will vary considerably, but in most cases a dedicated cadre of well-trained defense strategists is not likely to exist.

This is not to suggest that priority countries lack smart and capable individuals, but to point out that their systems struggle to identify those individuals, provide them with education and training on strategy formulation, and then assign them to the right functions to support strategy—and replace them with equally capable individuals when they subsequently are promoted, reassigned, or retire. In postauthoritarian countries, these challenges are compounded by a legacy of siloed institutional structures that prevented or discouraged the type of thinking strategy generation will require.

Tunisia is an example of a recently transitioned democratic state for which strategic thinking and strategy generation are largely new enterprises. Immediately after the fall of the Ben Ali regime, key individuals throughout the new government recognized that strategy generation would be a critical priority, but innumerable hurdles prevented quick action. The defense sector had largely been sidelined in favor of the internal security sector, and senior uniformed officers had generally been discouraged from engaging in strategy development. Absent also were the processes that would have facilitated the formulation of new strategy, and there was no dedicated cadre of officers or civilian personnel who were prepared for this important task. The challenge was compounded by a rapidly changed security environment, including a host of new threats for which the country was not well prepared. Despite these challenges, Tunisian defense sector personnel are well trained and educated and had benefited from opportunities to learn about strategy-making at defense and war colleges internationally. These individuals spearheaded several initiatives to begin developing institutions and processes for generating a national strategy and inculcating strategic thinking among senior officials and officers.

The Tunisian National Defense Institute (NDI) is the Tunisian Military's primary think tank, higher military education center, and the Ministry of Defense's official entity for defense strategy formulation.[10] Senior officials across the military ranks, including deputies, senior state clerks, and civilians from the three branches of Tunisia's military (Army, Air Force, and Navy), the Ministry of Interior, Customs, and Penitentiary Services are nominated to join NDI after completion of their courses at the Staff School and the Higher War School or following nomination by their individual ministry.[11] The Tunisian military annually sends its officers who are in line for promotion to complete a defense-focused course of study at NDI on a topic de jour. Often, this topic is strategic in nature. In 2016, as part of this effort to begin generating critical strategies, the class conducted studies to inform a National Strategy Against Terrorism and Violent Extremism,[12] and in 2017 that effort focused on creating a Tunisian National Security Strategy.[13] Various donors supported the NDI's efforts. For example, the United States helped arrange lectures and seminars in Tunisia and in the United States with strategy experts on how to develop a national strategy, and the Geneva Centre for Security Sector Governance hosted a conference with the Tunisian Armed Forces on how to develop a national defense strategy.[14] While the government ultimately never published the white paper that resulted from the 2017 effort, the efforts of the donors involved were importantly partner-driven. Donors did not write a strategy for Tunisia, but instead sought to support Tunisia's effort by providing technical expertise and guidance on the technical aspects of strategy generation.

In countries such as Tunisia, defense sector reform programmers could use a similar approach—working through existing training institutions to strengthen the content and focus of educational programs for strategy development. Alternatively, programmers could support the development of a dedicated program to generate a cadre of strategic thinkers who could support the development of a strategy generation system under the second principle. Other avenues for support could include sending strategic experts who could advise personnel assigned to the task of strategy generation, and in so doing, strengthen the defense sector's human capacity for strategy generation. Such an effort will likely require follow-on programming to introduce strategic thinking skills in existing curricula for midlevel and senior officers in the country's war colleges and senior service schools. Such curricula could, for instance, include lessons in strategic theory, the application of strategic theory to practice in real world scenarios, the evaluation of threats and opportunities in the current strategic environment, the civilian and interagency role in the strategy generation process, and frameworks for strategy formulation.

In countries with a more robust strategy generation process, including those that have received significant foreign assistance to develop their defense institutions, the existence of personnel who have received training or education in these types of strategy-focused skills will be more commonplace. For example, Georgia, which has received more than a decade of NATO assistance, has an Education and Military Training Command with several training and educational institutions that provide educational opportunities for members of the Georgian military, including courses that are necessary for promotion.[15] For example, the Davit Agmashenebeli National Defense Academy has a Junior Officer Training School and a Command and Staff College, which provide long-term, leadership-level professional education to military personnel.[16] In cases where higher-level professional military education is already established in the defense sector, an entry point for defense sector reform programmers may be to introduce more specialized education and training, ensuring that all essential tasks for strategy production (e.g., assessing the current operating environment, projecting likely future risks, analyzing what broad goals need to be achieved to successfully reach national security goals, or more broadly, how to write a strategy document) are being taught effectively and to the right people. Programming could also provide experts to assist the country in updating its existing training and education to include mandatory strategy courses for certain commander-level personnel (e.g., senior officers). More basic modules that explain the role and importance of strategy could also be developed for inclusion in the standard courses for lower levels to disseminate an understanding of and respect for strategy throughout the defense sector to encourage the broader cultural acceptance of its role in defense. If there are gaps in assigning personnel with the right skills to various strategy formulation roles, then additional programming may be required under *Goal 8: The Right People.*

2. A formal system (or systems) exists for the development and continuous adaptation and revision of the defense strategy.

Defense strategy generation requires that countries have formalized systems in place to *develop* a pragmatic national defense strategy; *articulate* the defense strategy on paper; *approve the strategy* through the correct channels; *communicate the strategy* effectively to relevant audiences; and *adapt* the defense strategy whenever necessary. Such systems ensure that the defense strategy remains current, appropriate, and relevant to the goals of the state and supports the objectives of the defense sector to mitigate risks and help ensure the military advantage. Although many countries in which defense sector reform will be a priority may purport to have a defense strategy, often there is no formal strategy generation system.

In countries that have been recently or are still mired in conflict, strategy may be ad hoc or nonexistent, and any strategy generation system that may once have existed will likely have given way to the necessity of redirecting all available resources to the immediate armed conflict. In such cases, if a strategy document does exist, it is likely to have little to no utility; without a generation process to keep the strategy updated and relevant, or a system to translate the strategy into plans, the document will quickly become outdated and useless. In Iraq, for example, the protracted 2003–11 conflict followed by a decade of continued instability has fully occupied every resource of the Iraqi defense sector and made the development of robust systems—like those necessary for defense strategy generation—a luxury that can rarely be afforded. Although attempts have been made to help mitigate the absence of strategy, the development of a strategy generation system has yet to occur. Iraq, for example, developed a National Security Strategy in 2015 with the assistance of the United Nations Development Program and the Geneva Centre for Security Sector Governance. This National Security Strategy, which was approved by the Iraqi National Security Council and the Council of Ministers, articulates the country's major threats and lists defense goals for Iraq. However, in practice, "it is in effect an ambitious, un-costed series of activities, without an associated implementation plan or road map."[17] The result is a complex document, prepared primarily by Western advisers, which does not accurately capture the challenges facing the Iraqi defense sector or Iraq's available resources.

In cases like this, the strategy generation system will likely have to be built from scratch, and determining where to focus initial activities may well be a complicated process. For example, in the case of Iraq, one of the major barriers to developing defense strategy lies in another separate but critical system: budgeting. Defense strategy must be tied to realistically available or acquirable resources, and Iraq's ability to budget needs significant development before an effective strategy generation is possible. The inability to effectively budget for future operations, coupled with the continued fight against ISIS, has put the Iraqi government in a reactive mode since 2014, and the result is haphazard strategy generation that is financially unrealistic, as well as being largely ineffective without massive support from outside countries.

In cases that require the comprehensive development of a new strategy generation system, one of the first hurdles for defense sector reform implementers will be deciding where to focus programming. For that reason, one good entry point is to conduct an assessment that can help determine, for example, what capabilities exist (e.g., some elements of the strategy generation cycle may be taking place in isolation from a broader coordinated system) or, if strategy documents have been developed in the recent past, whether they were ever implemented, and if not, why not (this line of inquiry may lead to clarity about other systems that need to be improved first, like the budgeting system in the case of Iraq).

Even in relatively institutionalized democracies, the strategy generation process—or indeed the existence of an overarching defense or security strategy—may be a point of weakness for the defense sector. For example, Tunisia is considered one of the most successful transitioning democracies in the region, yet although its security and defense forces are quite advanced and fairly well organized, the defense sector's strategy generation process is still nascent. Defense strategy in Tunisia is tied to the country's defense planning process, which the constitution delineates oversight responsibility for to the president. In reality, however, the president does not have the personnel available to coordinate the process; as a result, each of the armed forces develops its own plans without coordinating with the other branches. Although there are between 30 and 40 staff tasked with coordination in the Tunisian Ministry of Defense, there simply are not enough people to coordinate national-level efforts effectively. For example, within the president's office sits the Tunisian Institute for Strategic Studies (ITES). ITES operates as a think tank for the presidency, providing strategic analysis on security and development and offering training seminars and workshops for senior officials, often in tandem with international experts. Most of its members are retired diplomats. ITES could serve to assist with the strategy generation process, but it will need staff who have a high degree of knowledge of the defense sector and expertise on strategy generation.

This lack of a coordinated process is significantly more problematic due to the absence of an overarching national defense strategy to guide the armed forces toward unified objectives. As noted above, in 2017, donors worked with Tunisia to address this gap by assisting in the creation of a national security strategy, but the resulting white paper was never published and certainly never utilized. This may be due to the novelty of such a document within the Tunisian security sector. French officers who were part of the development process explained that the Tunisians involved in the strategy generation process saw it more as an academic analysis of the changing environment rather than as a blueprint for a strategy generation system.[18] Similarly, in 2016, Tunisia's National Security Council oversaw the development of a National Strategy Against Terrorism and Violent Extremism, and in 2017, began development of National Border Security Strategy.[19] Although both strategies were taken seriously by the National Security Council and the other agencies involved in their development (e.g., the Ministry of Finance, the Ministry of Interior, and the Ministry of Justice), neither strategy was ever published. The lack of single, unified defense strategy and of subsequent defense sector–wide substrategies, severely impacts the coordination and efficiency of defense planning for a country surrounded by active threats, not least of which is steadily increasing instability from its neighbors.

In such cases, a strategy generation system will likely need to be developed wholesale. Unlike the situation in Iraq, however, relatively established democracies will have more available structures and resources to support the process. Strategy experts could be made available to develop the series of processes that will constitute a strategy generation system, including designing processes that are currently missing or otherwise strengthening those that may already exist but are not well defined, coordinated, or staffed.

In other cases, particularly in advanced democracies, the system for generating strategy may be quite robust. Georgia has three national-level defense strategy documents that are codified in the Law of Georgia on Defense Planning: the National Security Concept, the National Threat Assessment, and the National Military Strategy.[20] The National Security Concept defines Georgia's core national values and interests; lays out the direction of the nation's security policy; and identifies existing or future threats, risks, and challenges. The document itself notes the involvement of a wide array of stakeholders in the strategy's development—"Political parties, nongovernmental organizations, and other representatives of civil society have played an important role in the drafting of the Concept"—and the strategy is generated by the government and ratified by Parliament.[21] The National Security Concept has been published only twice, initially in 2005 and again in 2011 in response to a perceived change in the environment due to aggression from Russia. The Ministry of Defense has referenced the ongoing development of a new National Defense Strategy, which will span the next decade through 2030.[22]

The National Security Concept serves as the overarching strategy that guides the development of other specific strategies. For example, Georgia also has a National Military Strategy, which defines activities and provides operational planning guidance to the Georgian Armed Forces; identifies military goals, objectives, and requirements; and provides guidance for the armed forces' structure and capabilities.[23] The National Military Strategy was most recently updated in 2014, but several subordinate strategic documents are published on a more frequent basis. For example, Georgia's Strategic Defense Review is published every four years (the most recent version was published in 2021 and covers the period 2021–25).[24] This strategic document "determines the structure and directions for the development of institutional and operational capabilities" of the Ministry of Defense and the Georgian Armed Forces.[25] The document also reviews the current security environment, threats, and challenges facing Georgia, while noting that it takes Georgia's "limited resources into the account and provides a set of time-phased and incremental development recommendations."[26] Since 2015, the Georgian ministry of defense has published an annual document called the "Minister's Directives" or the "Minister's Vision", which provides "a vision for the future with direction for today," for the development of the Georgian Armed Forces.[27]

Despite its relatively robust strategy generation architecture, Georgia faces a problem in terms of the frequency with which strategy is changed. While keeping strategy consistently updated to reflect fluctuating conditions on the ground is certainly important, frequent changes in leadership have resulted in these revisions being less updates and more wholesale reconceptualizations of the defense strategy and development goals. For example, the minister of defense is authorized to approve all concepts, strategies, and operational orders (including defense spending, organizational structure, and personnel) generated by the Ministry of Defense. This gives the minister significant influence over defining defense priorities and development priorities for the Georgian Armed Forces. However, almost every second year, a new defense minister takes charge, which often precipitates a significant change of strategic priorities, as can be seen in the annually published Minister's Directives documents.[28] The constant changes to defense strategy and plans fundamentally impede the continuity that is necessary for the long-term development of Georgia's defense sector and armed forces.[29] In cases where strategy documents exist but are revised and updated either too often or too infrequently, defense sector reform programmers may consider mapping the strategy generation cycle as an entry point. Having a clear picture of exactly when each strategy document is produced, revised, and replaced will help to determine what kind of programming can help the host country to optimize its strategy generation cycle.

CHAPTER TWELVE

GOAL 10
Military Effectiveness

The tenth goal of defense sector reform and institution building is *military effectiveness*. The military instrument translates the power and resources of the state to employ force for the functions of defense, deterrence, compellence, and show of force.[1] An effective military force is thus one that can, at a minimum, defend and deter attacks against the state and its population. More succinctly, "military effectiveness is the ability of a military force to successfully prosecute a variety of operations against a country's adversaries."[2]

At its core, military effectiveness is measured by the degree to which a country's leaders can *transform* national resources into warfighting capabilities that can be employed to "impose their will on enemies, existing and potential."[3] Merely having large armies or generous defense budgets does not equate to military effectiveness. Large defense budgets are an indicator of military strength, but how these funds are dispersed is more telling. Manpower, too, is a misleading indicator. Certainly, effectiveness is impacted by the size of the total force, the relative proportions of active and reserve forces, and the distribution of numbers among the services, but qualitative measures such as the educational and technical proficiency levels of both officer corps and enlisted ranks have a significant impact on effectiveness.[4] History is replete with examples of smaller armies that defeat larger ones because the smaller forces better transform resources into warfighting capabilities.

Physical infrastructure is also a component of military effectiveness. Here again, the number of installations is valuable as an indicator only to the extent those installations support warfighting capacity. Manicured headquarters facilities but rudimentary training installations for combat and combat support likely generate a hollow force. Yet another important contributor to military effectiveness is military inventory—not only the weapons and support capabilities but also the parts to repair them. And there are many other indicators, some of which may not be present in countries most likely to be recipients of defense sector reform efforts, such as the domestic industrial base for designing and producing military equipment, from low-level components to complex and technologically advanced systems.[5]

The key indicator of effectiveness is not how much a country has—in terms of budget size, manpower, physical infrastructure, military inventory, or domestic defense industry—but the ability to "convert" them into a "modern force capable of conducting effective operations against a wide range of adversaries."[6] The case of Iraq is illustrative. After years in which the United States tried to shape the Iraqi military into a force that looked like, fought like, and used the same equipment as the U.S. military, the ISF was rapidly defeated by a much smaller force of ISIS fighters in 2014.[7]

Although effective military force is the raison d'etre of the defense sector, it receives scant treatment in the literature on defense sector reform and defense institution building. The institution building literature recognizes that "the effectiveness and efficiency of defense, especially in times of peace, are very difficult to measure," but guidance for how to measure them highlights "aspects such as corruption or other clearly identifiable breaches of law"; it remains "an open field of discussion [about] what would be the most appropriate methods to assess the level of effectiveness and efficiency in defense."[8] Similarly, defense sector reform guidance acknowledges that "the overall objective . . . is to increase the ability of partner countries to meet the range of security and justice challenges they face." However, little attention is paid to defining effectiveness or providing even broad guidance for how reform will (re)build forces that can accomplish their missions effectively.[9] The lack of detailed guidance is surprising because such reform is often prompted by the *lack* of effectiveness, and military defeat is a powerful catalyst for reform. Cross-border incursions, terrorist attacks, the rapid militarization of a neighboring state, or the operational requirements of a new alliance can generate serious efforts to (re)build military forces, reform training, revise doctrine, or enhance professionalism and competency.

Indeed, the lack of effectiveness is often the entry point for such efforts. Reform initiatives may be generated initially by operational failure; improving operational effectiveness then prompts a closer look at institutional capacity for strategy generation, recruitment and force generation, logistics, procurement, and other essential defense sector functions. In other words, the ability to convert resources that generate military effectiveness depends on other key factors, such as the strategy generated to contend with the threats a country faces, the nature and structure of civil-military relations, and the ability to provision and deploy forces. As such, military effectiveness may be an entry point for defense sector reform, but successfully *achieving* the goal of military effectiveness will hinge on addressing the other goals of defense sector reform detailed in this guide. Military effectiveness is the sum of a complex equation involving essential and often interrelated processes across the defense sector. These include:

- An effective strategy generation process that links the state's security objectives to the military capabilities needed to address those objectives (discussed under *Goal 9: Strategy Generation*).

- A clearly defined relationship between civilian and military holders of power that defines the authorities for the creation and effective use of military force (*Goal 2: Civilian Control*).

- An acquisition system in which procurement decisions are based on strategic evaluations of what is necessary to conduct the missions the military is tasked with conducting (*Goal 7: Financial Management*).

- A professional officer corps that has been trained to execute decisions across a range of likely operational environments and consists of officers selected, retained, and promoted based on merit (*Goal 8: The Right People*).

- Functioning logistics systems for supplying and provisioning military forces to operate effectively (*Goal 5: Functioning Logistics*).

- Established mechanisms and processes for coordinating defense sector strategy generation and operational decision-making within and across the defense sector (*Goal 4: Management and Coordination*).

What guidance exists, in addition to the guidance under the other defense sector goals, for accomplishing the vital task of converting resources into military effectiveness?

Guiding Principles for the Design and Implementation of Goal 10

First, military institutions should have doctrine that specifies how a military uses its assets on the battlefield. Doctrine is essential for converting resources into combat capabilities because it guides the integration of technology and manpower to secure operational outcomes. Doctrine can be simply defined as "teachings," captured in fundamental principles that detail "how a military plans to fight."[10] Doctrine resides in policies and procedures for a particular military service or in tactics, techniques, and procedures for unit training and combat.[11] For some services, it is captured in manuals and standard operating procedures; for others, it is found in established tradition and transmitted through training, procedures, plans, and orders.

Doctrine derives from the analysis of past experience and an assessment of the needs and possibilities of the present and future.[12] As such, the process of generating doctrine is ongoing. New equipment and new experiences will necessitate revisions and amendments. Doctrine generation is not formulaic; it is a product of complex processes with different influences that will vary greatly to incorporate the evolving nature of weapons technology, the influence of formative experiences, organizational and institutional interests, national ideology and culture, and the institution's political and strategic environment.[13]

In many of the countries where defense sector reform is likely to be a priority or where operational failures promote such reform, doctrine is likely absent, dated, overly dogmatic, or divorced from strategy. In such instances, inferior or absent doctrine can negate the benefits of superior forces and equipment.[14] "Doctrine requires judgment in application," and once formulated, "it will have a continuous effect and impact on the routine operations of all forces."[15] Generating new doctrine to enhance military effectiveness will require not just the capture of teachings from past experience but also imaginative thinking about future needs of a force alongside and in support of a country's strategy. As such, the entry point for such efforts will likely begin with education.

In most established Western democracies, military institutions feature multiple required schools and training programs at the joint and service levels for officers of various ranks that teach the fundamentals of strategic thought and the history of warfare, among other subjects. Effective programs emphasize learning through application and analysis. In countries with less robust systems or that have no officer education programs, training often centers around rote learning. In such cases, defense sector reform efforts can support the creation or strengthening of educational offerings or build the knowledge and skills of host-nation military academy educators and trainers. Although improved education serves more than just the goal of military effectiveness, it may not be a necessity for generating new doctrine when such doctrine can be created by one or two individuals or when those individuals are not military officers. Additionally, new doctrine can also be generated from existing doctrine elsewhere. Furthermore, where doctrine exists but is not codified, interventions can focus on supporting that codification. Of course, the best teacher is often combat itself, and major doctrinal changes are usually generated by recent or current military operations. Rare is the organization that can generate new doctrine successfully in peacetime.

Often, the biggest challenge is not the generation of doctrine, but adherence to it once it is developed. Doctrine is only useful insofar as it is translated to generate value, which requires optimal command and coordinating structures, among other elements in the other defense sector goals detailed in this guide.

Second, military organizations should have optimal command and coordinating structures for the missions they are tasked with conducting. Organization is another essential element for converting resources into combat capabilities because it can support or inhibit operational outcomes. Military organizations with rigid command structures, highly bureaucratized or compartmentalized institutions, or siloed support functions are unlikely to produce the initiative and flexibility required to conduct operations, or to translate new doctrine into those operations, with maximum effectiveness.[16] Similarly, an officer corps that is chosen for regime loyalty or ethnic affiliation rather than operational effectiveness is unlikely to exhibit the initiative and agility to adapt and respond to conditions on the ground or to leverage opportunities effectively. Effective militaries exhibit a range of organizational structures, ranging from more centralized to more decentralized. The key issue is whether the chosen structure is optimal—the most appropriate—for the missions it is tasked to conduct. In other words, is the chosen organizational structure adaptable? Is it agile enough to respond to changing conditions, needs, and requirements, including technology? Does information flow effectively to enable sound and rapid decision-making and action?

In many of the countries where defense sector reform is likely to be a priority, the answers to these questions are often no. Likely candidates for reform are often countries with recent authoritarian pasts and military organizations that still reflect that past. In other countries, militaries may have served other (internal) functions, such as employment for key population groups or societal integration. Additionally, the military's main purpose may have been serving as the consumer of choice for local industries. In many cases, the institutions most in need of reform are bloated organizations with structures that are byzantine in their complexity. In still others, the organization may reflect colonial or other legacies unsuited to the present operational environment. In each of these instances, entry points for reform will revolve around right-sizing these structures and improving the flow of information and hence decision-making.

Third, military forces must undergo regular and iterative training—including exercises, war games, and other field-based training—to prepare for the missions they are tasked with conducting. Effective training is another essential element for converting manpower and equipment resources into combat capabilities. Military forces that receive poor or limited training will be unable to make effective use of their equipment no matter how well they are provisioned or how sophisticated their weaponry.

Effective training must encompass training for the individual and collective training for both large units (e.g., battle carrier groups) and small units (e.g., platoons or squads). It must also encompass conditions that are as realistic as possible and come as close as possible to the environment and situations faced in combat. In some cases, this training will include simulations and war games. Well-resourced democracies often have entire commands devoted to training, as well as robust and well-provisioned training centers that closely approximate field conditions in which large numbers of troops can train as a unit. In many of the countries where defense sector reform is likely to be a priority, there are limited opportunities for training and few resources to support realistic training conditions. Often, training facilities are poorly provisioned or entirely absent; material for training is in disrepair or in scant supply; and scenarios, exercises, and other training materials do not exist. However, training and the provision of equipment alone will have only limited impact if this capacity is not also institutionalized within the military to ensure that the training becomes iterative and can be sustained over the long term.

Fourth, militarily effective institutions must have a capacity and potential for innovation. The capacity and potential for innovation is difficult to define but nonetheless essential for the ability to convert resources into operational effectiveness. Innovation is what allows a force to adapt to changing strategic and operational conditions, develop solutions, and "stay one step ahead of its potential adversaries."[17] Indeed, this capacity is integral to the development of new doctrine, improved command structures, streamlined logistics, and enhanced training. Often, this capacity to innovate is the crucial factor that explains why a less well-equipped or smaller force prevails against a superior adversary.

But how is this capacity and potential for innovation achieved? One driver may be necessity. Military institutions that face serious, and even existential, threats are more likely to search for innovations because they have little choice. The cost of not doing so may be their very survival. But such cases are more likely to be the exception, and necessity provides little guidance for how to conduct defense sector reform to generate that capacity and potential for innovation.

One answer may lie in overcoming impediments to innovation. Military organizations tend to be conservative; in peacetime, their goals are often shaped by the need to maintain budgets, manpower, and operational domains. In such conditions, military organizations tend to focus on near-term, rather than long-term goals, and research is devoted to near-term problem solving rather than longer-term, and potentially disruptive, innovations. "These impediments to innovation are likely to be overcome . . . [when] organizations . . . [experience] failure," when they have "'slack' (that is, substantial uncommitted resources),” or when “civilian leadership intervenes to force military organizations to innovate.”[18]

Another answer may lie in the organization’s professional military and civilian leadership who, through selection, education, and promotion, feature reform-minded individuals or experienced leaders who have a vision about the future of warfare.[19] Here, the answer lies not in the removal of impediments but in the encouragement of analysis and forward thinking by a professional entity that is committed to its mission of providing security for the state. In such cases, entry points for defense sector reform may include generating new career paths for reform-minded officers or encouraging competition and debate within or between branches of the military. Given that many likely candidates for defense sector reform are low-tech countries with limited resources, the capacity and potential for innovation offers an asymmetric force multiplier. The ability to innovate may well enable that conversion capability so essential for military effectiveness.

Applying the Goal 10 Principles in Practice

These four principles provide conceptual guidance for how to achieve the goal of military effectiveness in a wide range of defense sector reform contexts. Their application in practice—particularly when designing defense sector reform activities or interventions—is a contextually driven exercise. Under each guiding principle, some examples of how these principles might generate activities or interventions are provided to help guide the practitioner to translate the principles into possible activities in a specific defense sector reform environment.

1. Doctrine specifies how the military uses its assets on the battlefield.

Doctrine is essential for converting resources into combat capabilities because it guides the integration of technology and manpower to secure operational outcomes. In many of the countries where defense sector reform is likely to be a priority, doctrine is likely absent, dated, overly dogmatic, or divorced from strategy.

The example of Tunisia is illustrative. Prior to 2011, Tunisia’s military was largely sidelined by a regime that feared a military coup.[20] The military’s operational experience was focused heavily on joint or cross-border operations with neighboring Algeria and Libya; operations in the desert to the south to monitor cross-border tribal movements and smugglers; and deployments overseas to UN peacekeeping missions, with the unintended effect that when Ben Ali’s regime fell, it was Tunisia’s military who “kept the peace.” Tunisia’s experience with peacekeeping operations overseas became the new "doctrine" for operations at home. Tanks guarded key installations and Tunisia's military forces were widely heralded as the "heroes of the revolution" for resisting orders to use force against Tunisian citizens.

With the collapse of the Ben Ali regime went the controls for monitoring cross-border movement, which had permitted the smuggling of otherwise legal goods, such as food, clothes, and fuel, and had involved cooperation with cross-border counterparts to prevent the movement of nefarious goods and actors. The intelligence network of agents along Tunisia's border also collapsed with the old regime. A range of new and more dangerous actors began entering and crossing Tunisia, including weapons and fighters from Libya bound for the Sahel.[21] Violent extremist groups launched attacks in cities and tourist centers and against critical installations and assassinated Tunisian politicians. Tunisian military forces suffered casualties against entrenched armed groups operating out of Mount Chambi along the western border with Algeria. ISIS successfully recruited foreign fighters in Tunisia, and between three thousand and six thousand Tunisians left to fight in Syria and Iraq. In only a few short years following its democratic transition, Tunisia faced an entirely changed operational environment—one that required new doctrine to capture the transformed operational requirements of Tunisia's military forces for new border security and counterterrorism missions.

The use and roles of the armed forces had to shift to account for "a complex and daunting landscape of threats and challenges, including counterterrorism, border security, and upgrading military professionalism and readiness."[22] With substantial technical and material assistance from donors, Tunisia began to shift its doctrine to account for the transformed operational and strategic environments, and by 2015 those efforts began to generate successes in counterterrorism and border security operations.[23] By 2016, the military was able to thwart an attack on police and military forces in Ben Guerdane on the Libyan border.[24]

Much of this has been accomplished with substantial and sustained security assistance, including out-year planning of needed equipment, budgeting for weapons system sustainment, and training of key units,[25] but there were also indications that efforts were being made to generate that needed innovation from within. For example, the Center for Military Studies has launched outreach to experts to introduce new teaching for the military, and the interagency National Defense Institute, akin to a senior-level war college, has established exchange programs with its counterpart in Washington, D.C., and each year's class is assigned a strategy initiative—such as drafting a defense white paper—as its major project.[26] However, one of the principal drivers of reform and adaptation in Tunisia, which suggests a valuable entry point for generating and translating new doctrine in similar contexts, is security cooperation with allied militaries and even joint operations. Wars, or in this case, security operations, are valuable drivers of change, particularly in defense sectors where innovation or flexibility has been previously constrained.

The emergence of new doctrine in Tunisia is still nascent—and only part of a larger effort to enhance military effectiveness. Tunisia's defense sector, like those in other transitioning authoritarian states that face serious security threats while undergoing regime shifts, lacks the infrastructure and resources that undergird doctrinal innovation. Legacies of underfunding, outmoded equipment, and overly bureaucratic institutions further impede the fundamental doctrinal shifts likely required, even though there is at least recognition that change is needed. In such contexts, likely entry points for defense sector reform programming could include sustained military-to-military advising and joint operations. Where needed, additional assistance can be provided by defense experts who could deploy to work alongside counterparts on particular initiatives, including the development of new doctrine or the codification of existing doctrine. Because such engagement will open these countries' defense sectors to outside scrutiny and will likely be treated with suspicion, care must be taken in selecting experts to ensure that the engagement is productive. And because trust will likely build slowly, such assistance should be designed for the longer term rather than through one- or two-week advising trips, which are unlikely to garner substantive engagement or produce meaningful change.

Another promising entry point is military education, and here the adage that more is better is in fact true. Security assistance programs that fund military officers to attend war colleges and other senior service schools overseas do more than enhance these officers' book learning.[27] They critically expose them to the workings of democratic militaries; introduce them to new and innovative ways of thinking and analysis, often lacking in their own institutions; and free them from the constraints of their own institutions, which in many recently transitioned states are still imbued with the practices of the old regime. These are often not places where "thinking outside the box" is encouraged, let alone safe to do. Overcoming that mindset is critical to modernizing doctrine and meeting the other goals of defense sector reform detailed in this guide, and exposure to other officers and institutions can be an effective way of changing mindsets. Where language fluency requirements are a hurdle and limit such opportunities to a select handful of officers, another option is to bring elements of this educational system to the country itself.[28] Although doing so may not re-create all the benefits of a year in residence overseas, creating or strengthening curricula or building the knowledge and skills of host-nation military academy educators and trainers may be critical to modernizing how and what the next generation of military officers or their senior leaders are taught.

2. Military organizations have optimal command and coordinating structures for the missions they are tasked with conducting.

Military organizations with rigid command structures, highly bureaucratized or compartmentalized institutions, or siloed support functions are unlikely to produce the initiative and flexibility required to conduct operations with maximum effectiveness.[29] Here, the example of Tunisia is again illustrative. Although Tunisia's military and its general staff are professional, apolitical, and subject to civilian oversight, long-standing tensions between the military and security forces, and between the ministries of interior and defense, constrain effective operational coordination between the two sectors. This is particularly noteworthy because the two leading security challenges Tunisia faces in the postauthoritarian period—counterterrorism and border security—require a high degree of cooperation and even joint operations under shared mandates.[30] Military command and coordinating structures are tied to zones of operations that predate Tunisia's democratic transition and the new security environment that emerged in its wake, with operational, territorial, and development units being deployed with varying degrees of experience in and knowledge of the regions to which they are assigned.[31] Often, National Guard units have better knowledge of the operational environment, but coordinating mechanisms between the National Guard and the army have been historically absent, although both have mandates for border security. Also lacking are command structures, particularly where multiple forces' operations overlap. "Coordination between the military and National Guard forces on the border remains tenuous, with some ad hoc coordination on the tactical level—joint patrols, some lower-level intelligence-sharing—but coordination remains a significant gap at the operational and strategic levels, particularly when it comes to clarity of jurisdiction and delineating areas of responsibility."[32]

Poorly structured or absent command and coordinating mechanisms in many of the countries where defense sector reform will likely be a priority suggests an important entry point for programmatic interventions. In some contexts, this will require streamlining such mechanisms, particularly where (as in many countries transitioning from authoritarian rule) such structures are large, cumbersome, or byzantine in complexity.

Officer selection for command may also require reform, not only in the context of meeting other defense sector goals detailed in this guide but also to achieve the goal of military effectiveness. In postconflict environments such as Libya or Iraq, reforming officer selection may require shifting selection process and promotion standards to prioritize operational effectiveness rather than regime loyalty or ethnic or tribal affiliation. In structuring assistance, the key criteria must be whether the organizational structure is optimal—that is, the most appropriate—for the missions the military is tasked to conduct. In other words, is the chosen organizational structure able to convert manpower and equipment resources effectively to conduct operations against its adversaries?

3. Military forces undergo regular and iterative training—including exercises, war games, and other field-based training—to prepare for the missions they are tasked with conducting.

Effective training is another essential element for converting manpower and equipment resources into combat capabilities. Military forces that receive poor or limited training will be unable to make effective use of their equipment, no matter how well they are provisioned or how sophisticated that weaponry. Training and multicountry exercises, including through alliances, are frequently a significant component of security assistance, alongside the provision of equipment. Translating this assistance into defense sector reform will require programmatic interventions to ensure that the training (and equipment, addressed separately under *Goal 5: Functioning Logistics*) can be sustained by the recipient defense sector institution after that training is provided or assistance ends. Entry points for reform in such instances can include supporting the development of training capacity; providing resources to enable realistic, field-based or scenario-based training; and supporting the better preparation of instructors.

Tunisia and Iraq are both examples of countries that have benefited from significant foreign assistance to train their forces for new operational mandates. For Tunisia, this assistance has included participating in military exercises with U.S., European, and regional militaries[33] and has enhanced Tunisia's operational capacity. In Iraq, this assistance was even more extensive, with U.S. forces mounting A3E (advise, assist, accompany, and enable) operations and providing air support, artillery support, medical evacuation, planning assistance, logistical support, and ISR (intelligence, surveillance, reconnaissance) assets.[34] When U.S. forces withdrew in 2011, however, little if any of this capability could be sustained.[35]

Beyond training through joint operations or multicountry exercises, defense sector reform interventions will need to focus on building host country capacity to deliver, at a minimum, effective training through simulations, exercises, war games, and other practical or operationally focused training. Where such capacity already exists, additional interventions could focus on building host country capacity to design such training, particularly as operational requirements evolve or new doctrine is developed. Well-resourced states often have entire commands devoted to training, as well as robust and well-provisioned training centers that closely approximate field conditions in which large numbers of forces can train as a unit. In many of the countries where defense sector reform is likely to be a priority, there are limited opportunities for training and few resources to support realistic training conditions. Often, training facilities are poorly provisioned or entirely absent; material for training is in disrepair or in scant supply; and scenarios, exercises, and other training materials do not exist. Likely programmatic activities could include addressing those gaps—building basic training facilities, providing surplus materials for exercises, or deploying experienced trainers to assist with training design or delivery. Even in rudimentary conditions, better or more realistic training scenarios or unit training may generate significant operational improvements. In the longer term, institutionalizing such training will require assisting host-nation training commands in establishing regular training for relevant forces, a task for which those commands may require assistance with planning, budgeting, and logistics.

4. Defense sector institutions have a capacity and potential for innovation.

Innovation is what allows a force to adapt to changing strategic and operational conditions, develop solutions, and stay ahead of its potential adversaries. Although the capacity for innovation is integral to the development of new doctrine, improved command structures, streamlined logistics, and enhanced training, it is extremely difficult to generate innovation, especially in peacetime, and even for the most resource-rich and advanced democracies. Innovation is often generated by research and development (R&D), the infrastructure for which is lacking in most of the countries where defense sector reform is likely to be a priority. Although from a conceptual standpoint, the capacity for innovation is critical for military effectiveness, developing that capacity is unlikely to be a realistic or practical priority for designing defense sector reform interventions.

A more realistic and implementable alternative may be to address impediments to innovation, and many of those impediments will likely be addressed by interventions aimed at achieving the other nine goals of defense sector reform detailed in this guide. A country's human capital is key to its capacity for innovation, and selecting, promoting, and educating "the right people" (*Goal 8*) may generate the commanders with the ability to convert resources into combat capabilities or the strategic thinkers who can generate effective strategy. The right human capital can also generate the institutional cultural changes that are critical for defense sectors undergoing democratic transitions.

Alongside addressing impediments to innovation, other entry points could focus on smaller-scale adaptations—that perhaps do not rise to the level of "innovation"—but nonetheless measurably improve military effectiveness. Unit, rather than individual, training, particularly if delivered under near-realistic operational conditions, could significantly improve the operational capacity of those forces, particularly where training has previously been largely rote, static, or classroom-based. Improved procurement processes and timely delivery of critical resources—such as ammunition—to deployed forces could also have an outsized impact on military effectiveness, as the example of Mali in the lead-up to the 2012 mutiny, attests. Identifying the right programmatic intervention will require a high degree of contextual knowledge to identify the critical procurement or logistical bottlenecks or training gaps to generate the type of intervention that can generate adaptions that in turn enhance effectiveness.

Another answer may lie in the organization's professional military and civilian leadership, particularly in terms of adaptations for their selection, education, or promotion. Here, the answer lies not in the removal of impediments but in the encouragement of analysis and forward thinking by a professional entity that is committed to its mission of providing security for the state. Entry points for defense sector reform may include generating new career paths for reform-minded officers or encouraging competition and debate within or between branches of the military. Given that many likely candidates for defense sector reform are low-tech countries with limited resources, adaptations that fall short of true innovation may nonetheless serve as an asymmetric force multiplier to generate that conversion capability so essential for military effectiveness.

CHAPTER THIRTEEN

TRANSLATING THE TEN GOALS OF DEFENSE SECTOR REFORM INTO EFFECTIVE PROGRAMMING

The purpose of this guide is to provide a practitioner-oriented conceptual road map to guide program managers and officers who are tasked with the monumental task of reforming defense sectors in a wide range of contexts around the globe. Donors active in the security assistance sphere have ample resources and program staff have wide latitude to design programmatic interventions, but generating real impact, or return on investment, for foreign assistance dollars remains a challenge.[1] Much attention has been paid to understanding why security assistance fails to achieve its intended impact, and numerous studies, programmatic innovations, and even new funding streams and government programs have sought to rectify the perennial challenge of translating donor funds and political will into sustained recipient action and change.

This guide has been developed to make a small but important contribution to this effort, not by resolving the perennial challenge (which is beyond the task of any manual), but by assisting the program managers and officers who are tasked with "doing something" to better define what they can or should seek to achieve and how to go about doing it. In the absence of clear guidance, programmers have often had to fend for themselves and have resorted to looking backward to discover "what we have done before." This tendency to look to the past for guidance for the future is understandable. Time frames are often compressed, funding cycles are short, and priorities are urgent. If a partner military is being overrun, there is no time to conduct an assessment or design an assistance program from scratch. Far easier is to look at a program or funding stream that has worked elsewhere and use it. But this expedient approach rarely causes us to pause and ask, Why? What is the purpose of this assistance? What is the goal we seek to achieve by providing this assistance? And if there is a more immediate or interim goal (e.g., "improve logistics"), how does that effort align with and support the achievement of the broader, long-term goal ("create an effective defense sector"), recognizing that improved logistics alone will not make the defense sector "effective"?

Here, a critical interjection is important. Not all security assistance aims at reform. Sometimes, donors provide material or training support to address a near-term, finite gap. The goal is not to reform how the organization conducts military operations, merely to help a particular unit at a particular moment in time operate more effectively. That is not reform, and the goals laid out in this guide are not meant for that kind of assistance.

But when the goal is *reform*—real, measurable, and sustainable change—it behooves us to stop and ask why and for what purpose these efforts are being implemented. If that assistance is meant not just for that one unit at a particular moment in time but to change how the entire force operates, then the challenge is a systemic one. Is the force (democratically) controlled? Is it subject to civilian oversight? Is it provisioned and manned appropriately? And so we can continue, posing questions prompted by each of the ten defense sector reform goals.

Each goal chapter identifies a critical element of that system for which change cannot be left to "what we have" and "what we have done before" in our portfolio of programs or funding streams. Rather, if reform is the objective, then we must begin our work by determining what reforms are needed and how to design programmatic interventions to address those needs.

Where to Start?

Not surprisingly, many of the goal chapters point implementers to an initial mapping or assessment as a starting point. Understanding what exists or how it functions is a logical first step in any program design. If the goal is improved logistics, for example, understanding what logistics system or processes exist is a necessary precondition for designing "improvements." That initial mapping is likely to be followed by a more comprehensive assessment, because knowing that a system exits but is inefficient will require process streaming that system to identify where and how improvements can be introduced. More complicated are contexts where the function is largely absent or so ad hoc that there is in effect no system. In such cases, mapping may be less useful. Instead, programmers need to understand how functions are being executed and, most importantly, why those functions exist and are being executed as they are.

In all cases, understanding the why is a critical first step. Inefficient systems may be deliberately so because they stem opportunities for corruption.[2] Elsewhere, inefficiency may stem from deliberate actions on the part of a key individual or unit because it enables corruption. In still other cases, the why may be a lack of resources, the lack of authority to make changes, or even simple inertia—because "this is the way it has always been done." Such an effort also identifies potential spoilers early in the process, enabling programmers to consider how to address their potential ability to stymie any change. Often, the answer to the question why? lies in legacies—inherited institutions, functions, or systems or previous donor assistance that was only partially implemented (as, for example, when a new system is introduced but the individuals who were trained how to use it have deployed elsewhere or retired). Legion are the stories of equipment being provided for which there are no spare parts and no knowledge of how to maintain it, resulting in "spare part lots" of broken equipment pilfered to keep the few remaining resources functioning. Following the 2012 Mali mutiny, warehouses of boxed supplies were uncovered that were never used to provision the front lines as forces were being overrun. Understanding the why is the first order of business when programming aims to improve what is broken or missing.

An alternate answer to the question of where to begin lies not in deciding which steps to take but in identifying the conceptual entry point. Each of the ten goal chapters provides a potential entry point, and the challenge for programmers is to determine where to start, particularly when the objective is nothing less than an "effective defense sector" but funding, time, and political will do not support addressing all goals at once. Countries in need of defense sector reform are likely to be clustered on the lower end of the capacity spectrum, each country having many needs, all of them apparently both important and urgent. Nonetheless, when programmers have been tasked to "do something," they need a place to begin.

A number of straightforward questions can help the programmer get started. Those questions begin by asking about the impetus that is driving this defense sector reform activity for the target recipient country. Subsequent questions—covered in later sections of this chapter—address sustainability, sequencing, signs of success, formal vs. informal systems, high- vs. low-tech solutions, and the target country's appetite for reform. Each raises important considerations that should be considered at the outset as programmatic interventions are being designed.

What Is the Impetus for Reform?

Questions about the impetus for reform get at the heart of what is driving the need for change, and the answers to those questions may point toward specific goal chapters that can help the programmer to decide where to begin "doing something."

1. Is the impetus to have the recipient country *start* doing something or *stop* doing something?

If the defense sector needs to *start* new behaviors, establish new or improved functions or controls, or adopt new practices, then the following goals may be places to start.

- The presence of nonstatutory forces, secret forces, secret prisons, or other off-book functions is likely a good indicator that *Goal 1: Democratic Control* is a starting point for reform. The absence of democratic accountability—meaning citizens have a say and a vote—is another clear indicator for starting with Goal 1. An effective defense sector should exercise control over the entirety of a country's defense sector and be accountable to its population.

- If the police dominate security functions and the military has been sidelined, then *Goal 1: Democratic Control* provides a starting point for empowering the defense sector under democratic oversight (and potentially counterbalancing the outsized role of the police).

- If the defense sector has been effectively destroyed by conflict or captured by militias or other armed groups, then the defense sector may have to be rebuilt from the ground up. A good starting point is to ensure that this rebuilding happens under democratic control, suggesting that *Goal 1: Democratic Control* is a place to start.

- If the chain of command does not, ultimately, reside with a civilian, then a place to begin is *Goal 2: Civilian Control.*

- If the defense sector functions like a black box, with little information about its most basic attributes, such as its budget, size, or operations, then *Goal 2: Civilian Control* offers a starting point for enhancing civilian participation or access to information, or for introducing much needed civilian expertise to inform strategy, planning, or policymaking. If lack of civilian oversight is a driver, then *Goal 3: Legislative and Judicial Oversight* may be a better entry point. Legislators need access to information about the defense sector if they are to execute appropriate oversight.

- If legislators lack information or expertise to execute their oversight functions, particularly if legislative review is a new process, then *Goal 3: Legislative and Judicial Oversight* offers guidance on how to build this new capacity.

- If a defense sector struggles to respond to crises or if unnecessary duplication of responses wastes resources, then there may be a requirement to build or improve existing mechanisms for coordination. If the gap is inter- or intraministerial, then *Goal 4: Management and Coordination* may offer a starting point. If the poor coordination results from absent, inefficient, or duplicative command and control functions, then *Goal 10: Military Effectiveness* may be a better place to start.

- If forces run out of bullets and food, or if their equipment remains in warehouses in the capital, then *Goal 5: Functioning Logistics* is a place to begin. This is also true if the preponderance of a military's equipment is sidelined for repair or if forces cannot be transported to the front lines when they are needed. And if the impetus is the lack not of troops and materiel, but of the planning to determine what is needed and when, then a better starting point may be *Goal 6: Defense Planning*.

- If recipient defense sector personnel present programmers with a wish list of equipment and training but cannot make a case for why it is needed or how it will be used, then a good starting place to build their planning capacity is with *Goal 6: Defense Planning*. If they have no plan for sustainment of that equipment or cannot identify what resources could be used to maintain and repair the equipment on their wish list, then a second likely starting point is *Goal 7: Financial Management*.

- The presence of largely inexperienced or newly appointed civilian and uniformed personnel following a conflict, coup, or democratic transition is a good indicator that *Goal 8: The Right People* is a place to begin. Goal 8 may also be indicated if a military cannot recruit or retain enough forces for its missions. And because having the right people is likely a requirement for every other defense sector goal, Goal 8 will likely guide programming for any defense sector reform initiative, even if it is not the starting point.

- The absence of a national security or defense strategy suggests that generating that capacity may be a good starting point. *Goal 9: Strategy Generation* offers a place to begin. Goal 9 is also a good entry point if there is a strategy but no capacity to operationalize it. Strategies are only effective if the words on paper can be translated into action.

- If doctrine is absent, not codified, outdated, or divorced from strategy, a place to begin is *Goal 10: Military Effectiveness*. The defense sector must be able to translate doctrine into operations; doctrine, like strategy, that sits on a shelf has little value. One way to ensure this translation happens is through effective, regular, and iterative training, as in the case of exercises, war games, and other field-based training so that forces are prepared for the missions they are tasked with conducting. More advanced defense sectors may benefit from steps to remove hurdles to innovation. Military institutions that face serious, even existential, threats are more likely to search for innovations because they have little choice. Goal 10 offers some suggestions for how to generate this capacity.

If impetus for the defense sector reform activity is to *stop* or change existing behaviors, functions, or practices, then the following goals suggest entry points for programming.

- If the defense sector is bloated, inefficient, or predatory, then *Goal 1: Democratic Control* may provide useful guidance for right-sizing activities to reduce or better distribute those powers. Effective defense sectors should neither dominate the other branches of government nor have the power to exploit legal mechanisms to achieve undemocratic results.

- If "exceptional measures," such as states of emergency, are the norm and not the exception, then *Goal 1: Democratic Control* may provide a useful starting point.

- If operational decisions or commitments are generated behind closed doors with no public scrutiny, or if defense sector officials can make decisions without checks and balances, then *Goal 1: Democratic Control* may provide a starting point to limit the prerogatives of the defense sector and subject it to civilian oversight.

- If defense sector officials refuse to attend legislative hearings, ignore summons for testimony, or refuse to share budgets, planned procurements, or other information required by law, then *Goal 3: Legislative and Judicial Oversight* may be a place to begin changing those practices. Goal 3 is also indicated when judges are being intimidated or when they cannot function without a security detail.

- If defense sector funds are mostly off-budget, set aside for government capture industries, or lost to corruption, a place to halt the siphoning of resources may be *Goal 7: Financial Management*. If the impetus is not about the financial functions per se, but about the control over financial decision making, then likely entry points are *Goal 1: Democratic Control* to limit those prerogatives or *Goal 3: Legislative and Judicial Oversight* to prevent the defense sector from circumventing that oversight.

- If promotion or assignment to key positions is determined by "who you know" rather than "what you have done" or what the force needs, then *Goal 8: The Right People* may be a place to stop generating a hollow force.

2. Is the impetus for reform driven by operational needs?

Most often, the requirement to "do something" is driven by operational needs. Those needs are often urgent, and programmers are expected to deliver fast results.

- If the impetus is to address command and control or ministerial decision-making, particularly when forces have shared operational mandates but their ministerial oversight bodies do not coordinate, then *Goal 6: Management and Coordination* may be a place to start. Postconflict and postauthoritarian states frequently lack these mechanisms, which hampers their ability to respond to threats and crises. If the poor coordination results from absent, inefficient, or duplicative command and control functions, then *Goal 10: Military Effectiveness* may be a better place to start.

- If frontline forces lack weapons, ammunition, or provisions, then *Goal 5: Functioning Logistics* is a place to begin. Goal 5 may also be a useful entry point if the assistance includes the provision of significant equipment or other supplies—particularly if the recipient defense sector does not have a reliable method to track this equipment or move it to where it is needed. If there is a supply chain in place but no ability to forecast and plan what will be needed when and where, then a better starting point may be *Goal 6: Defense Planning*.

- If forces lack training, or lack training for their specific operational requirement, then *Goal 10: Military Effectiveness* may be a place to begin.

3. Is the impetus for reform driven by alliance or partnership requirements?

From the programmer's perspective, this impetus may be the easiest to plan for, because the requirements are likely already clearly defined and enumerated. In fact, much of the defense sector reform literature is informed by the NATO PAP-DIB principles, which clearly delineate what countries seeking to join NATO need to do.

- Some of these requirements relate to the structure and purpose of the defense sector and have a strong normative component—notably, *Goal 1: Democratic Control, Goal 2: Civilian Control,* and *Goal 3: Legislative and Judicial Oversight.* NATO is an alliance of democratic states, and new entrants must meet that fundamental democratic requirement.

- But there is also a very practical element. Forces in alliance or operating in partnership must be able to operate together effectively. Likely entry points will relate to interoperability—in communications, in logistics, and operations. A good starting point is to review the list of requirements and then identify the goals that meet those requirements.

- There is also a likely time line for accession, with certain reforms being higher priorities than other reforms. Programmers can use these priorities to identify and sequence their assistance programming.

This does not mean that reform will be easier—only that from the programmer's perspective, there is at least clear guidance on what must be done and perhaps even clear direction on where to start.

Alliance or partnership as the impetus for programming has an added benefit that cannot be overstated. Membership is a powerful incentive because it offers real, tangible, and possibly near-term benefits. These benefits tend to be powerful motivations for embracing change.

4. Is the impetus for reform driven by poor use of resources?

Programmers may be asked to "do something" when countries have been recipients of ongoing assistance but have resources that are not being used or equipment that is not being repaired, or when countries are using resources inefficiently.

- If time is the wasted resource—because the defense sector cannot make decisions quickly to deploy forces, respond to an operational opportunity, or plan effectively—then *Goal 6: Management and Coordination* may be a place to address institutional efficiencies, while *Goal 10 Military Effectiveness* may be a place to tackle inefficiency caused by poor command and control.

- If equipment, transportation, provisions, or ammunition is used inefficiently or wasted, then *Goal 5: Functioning Logistics* is a place to begin. If the impetus is the lack not of materiel, but of the planning to determine what is needed and when, then a better starting point may be *Goal 6: Defense Planning*.

- If force numbers are inadequate, or if casualty rates are too high given the operational context, then *Goal 8: The Right People* may be the place to begin addressing the question of manning and assignment planning needs, whereas casualty rates stemming from poor training may require efforts under *Goal 10: Military Effectiveness* that focus either on operational command if poor decisions are the problem or on predeployment training if forces are not adept at delivering their missions.

- If the poor use of resources derives from inadequate or absent strategy, then the likely entry point is *Goal 9: Strategy Generation*.

5. Is the impetus for reform driven by undesirable behavior?

Programmers may be asked to "do something" in countries where there are clear normative or legal transgressions.

- If nonstatutory forces such as militias and other armed groups operate without official sanction—meaning their role and function is not officially approved or directed—then a place to begin establishing or reestablishing the defense sector's monopoly of force is *Goal 1: Democratic Control.*

- If forces operate outside the normal chain of command, are loyal to a politician or military commander rather than the government, control secret prisons or arms depots, violate military codes or rules of engagement, commit war crimes or human rights violations, or otherwise operate with impunity, then a few goals may offer places to start, depending on the nature of the impunity. *Goal 1: Democratic Control* is the place to begin curtailing this impunity in law or practice; *Goal 2: Civilian Control* to establish legitimate, accountable civilian authority; *Goal 3: Legislative and Judicial Oversight* to undergird that authority with effective oversight of authorities and budgets and to hold forces accountable to military codes of conduct; and *Goal 10: Military Effectiveness* to establish effective command and control over forces and limit their operational impunity.

- If the bad behavior stems from poor training or incompetent operational command, then *Goal 10: Military Effectiveness* provides guidance on establishing effective command and control and improving training, including predeployment training. If poor command stems from the selection of commanders for sectarian affiliations, political loyalty, or corruption rather than military competence, then *Goal 8: The Right People* may be a place to select, train, promote, and retain effective commanders.

- If corruption siphons resources from the defense sector, then the place to begin is *Goal 7: Resource Management* to establish appropriate financial and procurement controls. To ensure those controls have teeth, budgetary authorities may need to be limited by statute, suggesting that *Goal 1: Democratic Control* may also offer programming guidance. *Goal 3: Legislative and Judicial Oversight* suggests establishing oversight to ensure those controls are not circumvented and that officials are held accountable for the resources they control.

- If legislators fail to execute oversight or are complicit in the undesirable behavior, then subjecting that behavior to the light of day through *Goal 3: Legislative and Judicial Oversight* may be the place to begin.

Can It Be Sustained?

Transforming a defense sector, rather than merely providing assistance to a particular unit at a particular moment in time, requires that programming under any of the ten defense sector goals be sustained by the recipient defense sector after that programming ends. Merely stating that the goal is reform will not make it so. Programmers will need to consider the following three questions at the design stage to ensure that the recipient defense sector has sufficient resources to continue implementing the proposed reforms without donor assistance.

1. Are there sufficient resources to fund the proposed reform without donor assistance?

If resources exist but they are poorly managed, or if defense sector personnel cannot answer this question because they do not have reliable financial information (e.g., budgets are classified or budgetary and financial processes are absent), then programming under *Goal 7: Resource Management* may be a requirement alongside programming under any other defense sector reform goal. If the resources do not exist, then the proposed programming will need to be streamlined, simplified, or scrapped. A paper-based, analogue procurement system may be feasible, whereas a complex, computerized system may not be implementable because there is no funding to maintain the infrastructure for a digital platform.

2. Are there sufficient people—in terms of numbers and capacity—to implement and then sustain the proposed reform without ongoing donor technical guidance or advisory support?

This is a complex question because the answer is highly dependent on the nature and scope of the proposed programming. Generating strategy, developing doctrine, and fostering innovation may not require large numbers of people, but will place a premium on highly educated, innovative thinkers. Translating new doctrine into operations—by, for example, standing up a specialized combat unit—will require slightly larger numbers of people with specialized skill sets, which in turn will require associated systems to generate that unit (e.g., human resources, training, procurement). Building or rebuilding a defense sector from the ground up following conflict or as part of a democratic transition will require both large numbers of people and significant levels of expertise. Can the defense sector access pools of talent in the country? Is there a robust higher education system? Alternatively, are there defense sector schools that can train the right number of people with the needed skill sets? If not, then programming under *Goal 8: The Right People* may be required alongside any other defense sector reform goal to generate the human capital required. If the human capital gap is a large one, this may require the programmer to extend the program time line and sequence proposed reforms to enhance the likelihood of sustainability. Generating human capital is rarely a quick fix.

3. Are institutions mature enough to absorb the proposed reform and to continue those operations after programming assistance ends?

Many of the defense sector goals require establishing *systems* to execute defense sector functions more effectively, and systems require institutions to execute them. It is good practice to map what institutions exist and to assess their capacity to absorb new functions. If there is no time to conduct a robust assessment of that absorptive capacity, programmers can consider some of the following indicators: institutional size (in terms of personnel); quality of leadership; willingness of the leadership to adopt change; existence of basic processes and procedures (indicating that these are familiar and can be built upon); and available resources (primarily financial but also infrastructure, particularly if technology is involved). An institution that is effectively an empty shell will be able to absorb little, if any, change.

What Should Be the Sequence of Activities?

The answers to these questions about impetus and sustainability may have highlighted more than one entry point. Given that resources may not be available to initiate programming for each of those points, programmers will need to determine where they should target their efforts first and what should be the sequence of subsequent activities. This is a highly contextual exercise, and defense sector needs will have to be weighed against the priorities of the donor, its funding cycles, and assistance authorities. With these caveats in mind, the answers to the questions above do suggest some guidance that may be useful across all contexts.

- *Goal 1: Democratic Control* is a prerequisite for many of the other defense sector goals because it establishes the authorities and prerogatives of the defense sector. In postconflict and postauthoritarian countries, democratic control is the foundation for building or rebuilding the defense sector, the basis from which all other defense sector functions flow. It also sets the standards for starting—or stopping—defense sector behaviors or practices, creates the standards and enforcement authorities for addressing legal and normative transgressions, and is a critical component of combatting corruption or impunity.

- Many of the goals hinge on having the right people to implement new processes, execute new behaviors, adhere to new standards, operate effectively, and sustain new reforms. *Goal 8: The Right People* addresses how to generate the right people; *Goal 10: Military Effectiveness* focuses on effective training of forces and command and control; and *Goal 2: Civilian Participation* looks at the distribution of civilian expertise and authority across the defense sector. Because people are the linchpin of reform, focusing on enhancing the defense sector's human capital may need to come first.

- Sustainability is the key to return on investment, and ensuring that programming can be sustained will require investing in sustainment at the outset of any programming design. What specifically will ensure sustainability is highly context specific. Consider the example of a program to provision frontline forces more effectively. Such a program requires an effective logistics system, for which *Goal 5: Functioning Logistics* details the components. Such a system includes (1) generating the people, materiel, and services; (2) deploying personnel and materiel (and evacuating them for the purposes of maintenance, reconstitution, and medical care); (3) warehousing, storing, and maintaining the materiel; (4) sustaining the force until the mission is achieved; and (5) transporting materiel and personnel into, throughout, and out of a theater of operations. Each of these systems will have embedded systems within them that are integral to their successful execution and may need to be addressed at the outset. For example, generating people will require human resources functions, addressed in *Goal 8: The Right People,* and generating materiel will require procurement and budgeting functions addressed in *Goal 7: Resource Management.* Determining what needs to be provisioned and when will require effective planning, addressed in *Goal 6: Defense Planning.* The goal is not to overwhelm the programmer with an ever-expanding list of systems and functions but to highlight that some functional changes will have dependencies tied to capacities that are not strictly logistics. A perfectly designed logistics system cannot be sustained if there is no procurement mechanism or sufficient access to timely funding. Most systems will require highly skilled personnel, for which there will likely always be a human resource component.

- Incentive structures are also critical for sustainability and should be incorporated in the program design at the outset. A perfectly designed system may be implemented, but it is of no practical value if no one adheres to it. Although appropriate incentives will be highly context specific, they are likely tied to the behavior of people, suggesting that

Goal 8: The Right People is a place to start capturing those new behaviors in standards for recruitment, promotion, or assignment.

• Proper sequencing requires a sufficiently long time line, particularly if reforms are to be sustained. If funding cycles or authorities do not align with this longer time line, programmers should plan activities in phases, ensuring that the sequencing of those activities establishes the foundations in the first phase and builds on those foundations in subsequent phases. There is, of course, a risk in this approach: subsequent phases may not be approved for funding. Nonetheless, the time lines required by funding authorities are simply too short to generate real reform.

A Formal or an Informal System?

Assessment teams often discover when they first map or assess a defense sector slated for reform that there is no formal system for the specific function or capacity they are seeking to address. Absence of a system or capacity is itself a finding, but it is usually a rare one. More common is the discovery that while there is no formal system, key "functions" are nonetheless being executed in some fashion. These functions may be highly inefficient, largely ad hoc, or corrupt, but they do exist, and programmers will need to understand what those functions are and why they are being executed as they are before engaging in any program design.

Between the two extremes of a fully functioning, formal system and the complete absence of any system or function at all, there is a wide gray area. This is likely where most candidates for defense sector reform will fall. Clustered near the higher end of the capacity spectrum will be defense sectors that have some form of a system—offices and personnel are assigned functions, there are documented policies and processes they must adhere to, and there is some way to track and measure whether functions are being delivered, and delivered to some standard of effectiveness. But there are also likely bottlenecks, gaps, and poor performers within that system. The bigger or more critical the gaps, the further toward the lower end of the capacity spectrum the defense sector falls. Nonetheless, whether efficient or inefficient, there is a *formal* system in place. Designing a program to improve that system requires understanding how it works and identifying its key components and actors. Process mapping such systems will be easier because there is some documentation of how that system works or at least is supposed to work. Programming can then focus on the gaps identified by comparing what should be with what is.

What is far more complicated, when it comes to program design, is understanding and mapping *informal* systems. These, too, can fall across that gray area, from the higher-performing to the lower-capacity ends of the spectrum. These are still systems, because functions are being performed, to some greater or lesser extent, routinely, but they are informal because functions are not documented or mandated by law or statute. Well-performing informal systems tend to be those that function by the force of precedent and established practice. Ministerial leaders present their budgets annually to the legislature in September not because it is required by law, but "because this is how it has always been done." Similarly, there may be a three-step intake process for new recruits to the army, not because there is a standard procedure, but because the general has ordered that it be so. These systems can be highly efficient, although there are inherent risks. Experienced programmers will likely have many stories similar to this one: A particular unit or office operates very efficiently because there is a seasoned or experienced individual at its helm. But once that individual leaves and his staff is assigned elsewhere, one of the best-functioning offices becomes the worst. This is the risk of informal systems. Because processes and procedures are not documented, they depend for their continued existence on the individuals who champion, manage, and abide by them. When those individuals are promoted, reassigned, dismissed, retired, or killed in action, the knowledge goes with them. It is for this reason that many of the defense sector goals identify formal systems as a requirement.

Informal systems are far more difficult and time intensive to understand and process stream. Because documents do not exist, information must be gleaned by talking to people and mapping the process as individuals reveal how it is done and by whom. Designing a program to improve this system—and formalize it—raises other challenges, including the likelihood that current users might well resist change. If it is working, why should they change it? Change is hard, and learning and implementing a new system requires more work than continuing to use the existing system. Here, quick wins may be important to generate support, and a clear explanation of what benefits a formal system may bring can do much to sustain that support for the duration of the implementation period.

Sometimes, a series of ad hoc processes can be mistaken for an informal system, particularly when those ad hoc approaches seem, for some inexplicable reason, to be working or at least to be performing some basic functions. How can a programmer tell the difference? Here, the test is the following: "If they can't explain it, you probably can't either."

Consider the following scenario. A team is sent to map what the programming team believes is an informal system. They interview the head of the unit, who explains very definitively that the process works this way. A subordinate describes a very different process. Staff who apparently implement the day-to-day functions of the process have never heard of it and cannot recall ever being told or required to implement any of the steps. Subsequent interviews with units on the generating and receiving ends of the process similarly describe something very different, or have no knowledge of this process at all. Except for the boss who answered definitively, few people involved in the process can explain how it happens. They provide examples of how it worked this time or that, but there is no consistency or routine. It seems to happen differently every time. It is clear no one understands the process they are part of, and the assessment team cannot map anything because there really is no process to map. This is not an informal system but a series of ad hoc tasks delivered by individuals as they see fit or as they are ordered to do.

In such instances, programmers will likely need to design a new system from scratch. Before they do so, it will be worth trying to understand why this ad hoc system exists, rather than an informal one. Is it a lack of resources, a lack of human capacity, or something else entirely? Knowing why will be important because it may identify some potential spoilers or pitfalls that could stymie implementation of a new system.

A final note of caution is in order. Sometimes, defense sector counterparts will withhold information about how the process really works, claiming they do not know or cannot explain it when in fact they do and can but choose not to. This is not uncommon, and it is even understandable. Donors should remember that these counterparts are often professionals who have dedicated their careers to the defense sector. They are likely proud of what they have been able to do, often with limited resources and under difficult conditions. They may well resent donors digging into all the reasons the organization is not working as well as it should. In other cases, they oppose the proposed reform or the politician that is promoting it, or they stand to lose from process improvements that will impact them personally, particularly if it threatens loss of power or income. It thus behooves the program design team—and their on-the-ground assessors—to cast their net widely when attempting to understand a process and the reasons why it does or does not work well. The more sources of information, the more likely the design team will have a better understanding of the system or function they seek to improve and the hurdles or roadblocks that might prevent successful implementation.

A High- or Low-Tech Solution?

It is not surprising that programmers—or their assessment teams—will recommend activities that draw on systems or processes with which they are familiar and know to be successful. When time lines are compressed and needs are urgent, a proven solution may seem a wiser choice than attempting to implement something new. But in many of the environments where defense sector reform is likely to be a priority, what works best is not a cutting-edge, high-tech solution, but something that is paper-based, analogue, or low- or no-tech. In other words, something "old" may be preferrable over something new because it is implementable.

The literature on security assistance offers many examples of technology-intensive reforms that fail because the most basic infrastructure—what many in the donor community may take for granted—is missing. Consider this example. There is a border post a dusty daylong drive from the capital. It is a rustic concrete block structure. It has no connection to the internet. In fact, it does not even have electricity. There is a generator, but it has not been maintained in years, and reliable fuel to power it is much desired but rarely available. Yet this outpost sits on one of the major trade routes, one that is increasingly being exploited by nefarious actors moving goods and people. Donors are concerned about the military's inability to track this movement and is offering to assist the ministry to implement a border management system. One proposed solution is an off-the-shelf integrated computerized system linking this outpost to headquarters with a central database to record and furnish the government with real time data. The scoping team that reviewed the proposed program design visited a border outpost selected by the defense ministry; because of the team's tight schedule, it was close to the capital and had a reliable electricity supply. But for every functioning border post there is another that is not. The implementation team then deploys and realizes that key outposts cannot implement the system. The high-tech solution for systemwide improvement can be implemented only for part of that system. Of course, there are some work-arounds—phone lines, a fax machine, a paper tracker of cross-border movements, staff at headquarters to upload the information into the system, and so on—but these have to be made ad hoc for each post. And because the "system" varies based on location, the institution now needs to develop different procedures for each location and different trainings for personnel based on where they are deployed, which now requires different guidance for human resources personnel. Given the costs, the added administrative requirements, and the likely challenges to sustainment, is the new solution worth the investment? The answer hinges on whether the new system generates data that improves border operations and, ultimately, if those improvements can be sustained. The answer is typically no.

In other cases, it is not just the infrastructure that presents challenges. Human capacity may also be severely limited. There may be few staff with sufficient computer skills to work the new system. Even a paper-based system for tracking spare parts can be a challenge to implement because personnel at the proposed logistics depot lack numeracy skills.

This high- vs. low-tech solution problem is not just an issue for countries on the lower end of the capacity spectrum. High-tech solutions are often developed for a particular context—an institution, its operational requirements, its existing system, or its resources—and these parameters do not necessarily translate to other contexts. Sometimes, higher-tech solutions offer the right fix, but they should not be the default solution.

Programmers should consider available infrastructure, human capital requirements, and, most importantly, how the existing system functions to determine if the proposed solution will work with that system. In so doing, a simple analogue approach may be the right choice because it can be implemented at the least cost and has the highest chance of being sustained.

How to Resist the Lure of Numbers?

Another pitfall to avoid in designing programming for defense sector reform is to focus exclusively on activities that generate big numbers and fast results. It is not surprising that provisioning equipment is often preferred over reforming an institutional process because equipment can easily be counted whereas process change is hard to measure. When program managers need to report results to sustain their funding at the one-year mark, one thousand items provided will appear a bigger win than one deployed expert to design a new system. At the two-year mark, that disparity might be even greater. And because reforming a system takes time, it may be yet another year (or even another five or ten years) before the first results of that new system can be identified. And identifying those results takes much more time and effort than counting items.

Often, program managers recognize that providing items over implementing process changes will not result in a more effective defense sector, but they do so anyway because funding cycles are short and the absence of results risks a reduction or cancellation of their program. Frequently, their approach is to register some reportable wins while implementing the longer-term, systemic changes that real reform often requires. This pragmatic approach is often dictated by the short-term nature of funding authorities for programs that realistically might take a decade or more to achieve their intended change.

The caution here is to recognize that the provision of equipment, or the training of large numbers of personnel without linking that training to institutional changes, is a shorter-term activity that may generate valuable quick wins but is unlikely to lead to sustainable change. If the goal is sustainable change, then activities that generate "big numbers" must be accompanied by more fundamental reforms. "Big number" activities are no substitute for real reform.

How to Tell If It Is Succeeding?

As projects are launched and implementation continues, programmers will have to answer the "success" question countless times, most importantly when they request an extension or expansion of funding or when they need to make the difficult decision to end a project because it has failed to achieve reform objectives.

Many program managers can rely on specialists to design a monitoring and evaluation plan that identifies indicators of success; collect and analyze data against those indicators; and conduct periodic evaluations throughout the program cycle. Increasingly, donors are recognizing the value of monitoring and evaluation and building teams of experts to ensure programs can answer the question of success. However, if programmers do not have access to this expertise, they may still need to answer the question. It is good practice to consider what success looks like at the outset as programmatic interventions are being designed.

The following five prompts may help the programmer to identify what success looks like and how to convey it.

1. Is work being done?

This may be an obvious question, but it is a highly revealing one. And "work" does not mean countless meetings without results. Work means that tangible products (e.g., a new procedure or process, a new policy or set of guidelines, a new tracking tool for supplies, a new draft statute or constitutional amendment) are being produced, not just talked about or promised. The seventh meeting of a ministry working group that continued to show promising agreement on the need for improving recruitment is not work. The creation of a draft procedure for recruitment is.

2. Who is doing the work?

The answer to this question is another important indicator of success, because if all the work is being done by a donor's advisory team, it indicates that the donor may want the outcome more than the recipient does. But if the work is being done by recipient defense sector personnel, then that is an indicator of success. It means that the institution values the proposed work, is putting resources behind that work, and potentially views the product of that work as its own.

3. Who is claiming ownership of the work?

If senior defense sector officials claim the work or the products as their own, that is another indicator of success. It demonstrates that the defense leaders value the work their subordinates are doing and see benefit in championing it and even claiming it as their own.

4. Is there formal validation or adoption?

If work products are formally validated by senior decision-makers or other government bodies, or if they are promulgated or adopted by the organization, this is a strong indicator that the recipient defense sector is not only doing the work, but also institutionalizing it. Adoption is a particularly important indicator of success because it will require the organization to adhere to the new requirement, which is key to sustainment.

5. Are products of the work iterative, routine, or ongoing?

Once the work is promulgated or adopted, a good indicator of success is if it generates ongoing activities, products, or work. Too often, new strategies, manuals, procedures, laws, or other products are produced and collect dust on a shelf or are otherwise ignored. The real test is if these generate the sought-after changes. For instance, if a new law requiring legislative oversight of the budget is followed by a legislative hearing to discuss proposed expenditures for the next fiscal year, that indicates success. If business continues as usual, then success has not been achieved.

There are, of course, many more sophisticated measures of success. But if programmers need to defend their program or advocate for new or sustained funding, then products, ownership, adoption, and adherence are good places to start.

"We Can't Want It More Than They Do"

One the biggest hurdles to defense sector reform is generating the political will to start and then the political commitment to see it through.[3] This kind of reform cannot be imposed. A program officer can attempt to identify and understand the problems and needs of the target country, design innovative and effective assistance to help the country address those reforms, and make generous resources available to support the effort, but none of this will result in more than superficial change or perhaps temporary improvements if the political will is not there among recipient defense sector institution personnel and, ultimately, their political leadership.

Donors support defense sector reform activities because it generates benefits to the donor or serves national interest. But donor benefits do not necessarily translate into recipient benefits. Often, recipient defense sectors will welcome the provision of expertise or equipment but resist implementing the changes necessary for that new expertise or equipment to have the impact the donor wants. Sometimes, that resistance manifests as slowing down the pace of reform in the hope that the donor will tire of pressing for change. In other instances, reform is actively opposed or undermined by elements within the government or the defense sector because it threatens the systems by which they rose to power, maintain their influence, or enrich themselves.

Activities under some goals have a high likelihood of generating resistance. *Goal 1: Democratic Control* will, by definition, disenfranchise those who have benefited from the nondemocratic structures that this reform seeks to replace. But even seemingly less political or more technical reforms can provoke resistance. For example, growing the size of the force to better meet a country's security challenges may seem like a goal that would generate wide support. But if that reform requires shifting recruitment targets from a privileged to a wider population group, and in so doing dismantles senior leaders' patronage network, then a seemingly innocuous reform becomes a highly political one. These "technical" or "process" reforms can generate reform-ending resistance because they are less likely to generate the public scrutiny and support that new constitutional provisions or laws might inspire.

Reforms mandated by partnership or alliance requirements are one noteworthy exception. They offer real, tangible, and often near-term benefits that will significantly reduce the likelihood of sustained resistance. Political leaders can use these promised benefits to generate support for change—even when that change disempowers key elites—because the benefits are clear-cut, as are the costs of refusing to implement reform. And the implementing staff—those who will have to do the hard, day-to-day work that change normally requires—will be less likely to actively oppose or delay the change because it promises real results. Put simply, it is easier to generate political support for fundamental change that promises concrete rather than uncertain benefits that may not manifest for a decade or more.

But in most other instances, programmers should anticipate that reform activities will generate opposition and may even inspire deliberate efforts to undermine the change. One way to mitigate this risk is to identify potential "change agents"—political leaders or defense sector personnel who support the reform—and help these change agents make their case and create buy-in from stakeholders within and outside the defense sector by prioritizing programming to yield quick wins. Quick wins can result from activities that are easy to do (because they encounter little resistance or require no fundamental change) or provide much-needed resources, or for which there is already significant baseline capacity to absorb the proposed change. The benefit of a quick win is that it empowers change agents to demonstrate the concrete benefits of reform to overcome potential resistance to other, longer-term or more fundamental changes that follow.

A note of caution about "change agents" is required. The term is widely used, but real change agents are few in number. They are the individuals who are in key positions to effect change, who believe change is both urgent and important, and who are willing to risk action because of the value it brings to the institution, not just themselves. To further empower these change agents, programmers may want to consider implementing early activities to generate like-minded colleagues in the same unit or organization or in other units or organizations that may serve as linchpins for the proposed reform or could stymie its successful implementation. Such activities will be highly context specific. They could include providing specialized training in a sought-after or career-enhancing subject, offering an embedded or shadowing deployment at a donor ministry or command, or providing equipment that the change agent could help allocate or formally receive on behalf of the organization. Opportunities for training at donor war colleges and senior service schools may also generate individuals who can, in time, better translate the benefits of reform to their colleagues.

No *reform* programming—no matter how perfectly designed or how generously funded—can substitute for recipient agency. Where recipients are willing to receive but not to act, programmers may be forced to scale back their plans so as to "do something—but at a smaller scale" or to to "do something—but at a later date," when recipient needs or conditions propel a desire for change. Ultimately, donors cannot value the change more than the recipient organization does. Reform must be valued, owned, and driven by the recipient political or defense sector leadership for it to be successful and sustained.

A Few Closing Thoughts

Practitioners know firsthand that the environments for which they are being tasked to design defense sector reform programming are considerably more complex than those the existing literature was meant to address. Most existing guidance emerged out of the successful experiences of Eastern European countries joining NATO at a time when the environment was more binary and stable than it is today. But the end of the Cold War and the far more complex twenty-first-century security environment that has emerged in its stead have vastly complicated the task of the defense sector reform programmer. There are many more unknown actors and forces operating in new ways in this environment. When programmers are tasked "to do something," it is often because defense sectors are on the losing end of battles against powerful nonstate actors. In other cases, this assistance is needed because donors anticipate they may need countries' help to battle such forces in critical parts of the world. The traditional mechanisms featured in that existing literature no longer apply to this more complex and less stable environment, and programmers need revised guidance that fits this new context. That need has inspired this guide, which has been developed after talking with defense sector reform practitioners in the United States and Europe, as well as those deployed in countries where such work is currently being conducted.

We learned from those discussions that one of the biggest hurdles is determining *what to do*—and more specifically, *where to start*. We have thus closed this guide with a final chapter that focuses specifically on the difficult questions programmers often face at the outset about the impetus for the reform, as well as other initial questions about sequencing and design that can shape the entire trajectory of that assistance. Although they are important at the beginning, these same considerations appear throughout the program design cycle. Defense sector reform programming is never a static exercise. New recipient or donor staff may accelerate or slow the pace of programming, or a shift in the security context may add additional urgency to do something, generating a new impetus for program design.

As we have argued throughout this guide, the art of the possible in program design is largely determined by context. From year to year, some of the factors that shape that context, such as historical legacy or geographic location, remain mostly fixed, but others, such as a country's social or political dimensions, can shift quickly. These shifts may expose new gaps or needs, generating a fresh set of demand signals that either open the door to reforms that may not have been possible at the outset or close the door to transformative change envisioned in the original program design.

At such junctures, programmers may well turn to this guide again because the ten defense sector reform goals serve as a menu or checklist of what can be done if that door is opened, as well as guidance for what do when it closes. The ten goal chapters provide guidance for implementation across a range of country contexts, and this final chapter offers suggestions for how programming can be scaled and sequenced if demand signals suggest there is an opening for transformative change. The chapter also provides a possible roadmap for quick wins and shorter-term programming options when there is no such opening. Empowered by the ten goals of defense sector reform, practitioners can pivot to leverage potential openings or set the stage to generate those openings in the future.

NOTES

Chapter 1. The Need for a Practitioner's Guide

1. Ashley J. Tellis, Janice Bially, Christopher Layne, and Melissa McPherson, *Measuring National Power in the Postindustrial Age* (Santa Monica, CA: RAND Corporation, 2000), 133, https://www.rand.org/pubs/mono graph_reports/MR1110.html.

2. "Process streaming" is a tool used in institution building and in business that involves diagramming a process using process streaming software or just pen and paper to discover if and how it works and how it can be streamlined or otherwise improved.

Chapter 2. Adapting Existing Guidance for Future Defense Sector Reform

1. United Nations, Department of Field Support, *Policy: Defense Sector Reform* (New York: United Nations Department of Peacekeeping Operations, 2011), 16, https://issat.dcaf.ch/download/18534/216935/UN%20 Defence_Sector_Reform_Policy.pdf.

2. Alexandra Kerr, "Introduction," in *Effective, Legitimate, Secure: Insights for Defense Institution Building*, ed. Alexandra Kerr and Michael Miklaucic, ix–xxvii (Washington, DC: NDU Press, 2015), xv.

3. The defense sector is a component of the larger security sector, and as such, the concept of defense sector reform falls within the parent concept of SSR. SSR can be broadly defined as "the political and technical process of improving state and human security by making security provision, management and oversight more effective and more accountable, within a framework of democratic civilian control, rule of law, and respect for human rights." SSR encompasses reform of all the major components of the security sector, including the internal security sector (e.g., the police, the ministry of interior), the justice sector (e.g., the court system, the ministry of justice), and the external defense sector (e.g., the armed forces, the ministry of defense) and includes reforms ranging from "legislative initiatives; policy making; awareness-raising and public information campaigns" to "management and administrative capacity building; infrastructure develop-ment; and improved training and equipment." See Geneva Centre for Security Sector Governance (DCAF), "SSR," https://www.dcaf.ch/about-ssgr.

4. Alix Julia Boucher, *Defence Sector Reform: A Note on Current Practice*, Stimson Future of Peace Opera-tions Program (Washington, DC: Henry L. Stimson Center, 2009).

5. Office of the Secretary of Defense for Policy, "DOD Directive 5205.82 Defense Institution Building (DIB)"U.S. Department of Defense, May 7, 2017, https://www.esd.whs.mil/Portals/54/Documents/DD /issuances/dodd/520582p.pdf?ver=2019-02-04-144847-587.

6. Kerr, "Introduction," xv.

7. The original handbook was updated in 2008 to include chapters on measuring and evaluation and gender mainstreaming based on feedback collected from OECD member states.

8. Organisation for Economic Co-operation and Development (OECD), *OECD DAC Handbook on Security System Reform: Supporting Security and Justice* (Paris: OECD, 2007), 3, https://read.oecd-ilibrary .org/development/the-oecd-dac-handbook-on-security-system-reform_9789264027862-en#page1.

9. Ibid., 21.

10. Ibid., 124–139.

11. Ibid., 124.

12. Ibid., 128–133.

13. UN, Department of Field Support, *Policy: Defence Sector Reform*, 1.

14. Ibid., 1, 3.

15. Ibid., 4.

16. Ibid., 3.

17. Ibid., 5.

18. Office of the Secretary of Defense for Policy, "DOD Directive 5205.82 Defense Institution Building (DIB)," January 27, 2016. Minor changes were made to the DIB Directive in May 2017; that version is available online at https://www.esd.whs.mil/Portals/54/Documents/DD/issuances/dodd/520582p.pdf?ver=2019-02-04-144847-587.

19. Ibid.

20. Ibid., 13. Note that "security cooperation" is a Department of Defense term for the "security assistance" activities it carries out.

21. Ibid., 4.

22. Ibid., 4–5.

23. For details of the Partnership Action Plan, see Philipp H. Fluri and Eden Cole, eds., *Defence Institution Building: 2005 Partnership Action Plan on Defence Institution Building Regional Conference* (Vienna: Austrian National Defence Academy; Geneva, Centre for Democratic Control of Armed Forces, 2005).

24. North Atlantic Treaty Organization (NATO), "Partnership Action Plan on Defence Institution Building" NATO, 2004, https://www.nato.int/cps/en/natohq/official_texts_21014.htm.

25. Ibid.

26. Wim F. Van Eekelen and Philipp H. Fluri, eds., *Defence Institution Building: A Sourcebook in Support of the Partnership Action Plan (PAP-DIB)* (Vienna: Austrian National Defence Academy; Geneva, Centre for Democratic Control of Armed Forces, 2006).

27. Ibid., 14.

28. "Partnership for Peace: Framework Document," Ministerial Meeting of the North Atlantic Council/ North Atlantic Cooperation Council, NATO Headquarters, Brussels, January 10–11, 1994, http://www.nato.int/docu/comm/49-95/c940110b.htm; and NATO, *Basic Document of the Euro-Atlantic Partnership Council* (NATO: Brussels, May 30, 1992), http://www.nato.int/cps/bu/natohq/official_texts_25471.htm?mode=press release.

29. Organisation for Security and Co-operation in Europe (OSCE), *Code of Conduct on Politico-Military Aspects of Security* (Vienna: OSCE, December 3, 1994), http://www.osce.org/fsc/41355.

30. Fluri and Cole, eds., *Defence Institution Building*, 2.

31. See, for example, Hari Bucur-Marcu, *Defence Institution Building Self-Assessment Kit: A Diagnostic Tool for Nations Building Defence Institutions* (Geneva: Geneva Centre for Security Sector Governance [DCAF], 2010).

32. Wim F. Van Eekelen, "What Kind of Defense Do We Need?" in *Defence Institution Building: A Sourcebook*, 4–18, 10.

33. Ibid.

34. Robert M. Perito and Jamie Kraut, "SSR in Libya: A Case of Reform in a Postconflict Environment," in *Prioritizing Security Sector Reform: A New U.S. Approach*, ed. Querine Hanlon and Richard Shultz, 37–66 (Washington, DC: USIP Press, 2016); and Querine Hanlon, "SSR in Tunisia: A Case of Postauthoritarian Transition," in *Prioritizing Security Sector Reform*, 67–93.

Chapter 3. Goal 1: Democratic Control

1. Hans Born, "Democratic Control of Defence Activities," in *Defence Institution Building: A Sourcebook in Support of the Partnership Action Plan (PAP-DIB)*, ed. Wim F. Van Eekelen and Philipp H. Fluri, 78–107 (Vienna: Austrian National Defence Academy; Geneva: Geneva Centre for the Democratic Control of Armed Forces, 2006), 78.

2. Fred Schreier, "The Division of Labour in the Defence and Security Sphere," in *Defence Institution Building: A Sourcebook*, 19–77, 23.

3. Ibid.

4. See Melanne A. Civic and Michael Miklaucic, "Introduction: The State and the Use of Force: Monopoly and Legitimacy," in *Monopoly of Force: The Nexus of DDR and SSR*, ed. Melanne A. Civic and Michael Miklaucic, xv–xxv (Washington, DC: National Defense University Press, 2011); Herbert Wulf, "Challenging the Weberian Concept of the State: The Future of the Monopoly of Violence," Australian Centre for Peace and Conflict Studies Occasional Paper, no. 9 (December 2007); and Martina Fischer and Beatrix Schmelzle, eds., *Building Peace in the Absence of States: Challenging the Discourse on State Failure* (Berlin: Berghof Research Center, 2009).

5. Civic and Miklaucic, "Introduction: The State and the Use of Force," xvi.

6. Schreier, "Division of Labour," 25.

7. Wulf, "Challenging the Weberian Concept of the State," 10.

8. Ibid., 11.

9. Ibid., 28.

10. Born, "Democratic Control of Defence Activities," 79.

11. Ibid., 81.

12. OSCE, "1994 Code of Conduct on Politico-Military Aspects of Security," December 3, 1994, https://www.osce.org/fsc/41355?download=true.

13. Schreier, "Division of Labour," 24.

14. Born, "Democratic Control of Defence Activities," 78–79.

15. Schreier, "Division of Labour," 32.

16. Ibid., 26.

17. Ibid., 45.

18. Ibid., 44.

19. Ibid., 44–45.

20. For example, a report in 2020 disclosed the existence of a secret prison in the Tahrawa area in the Nineveh Governorate of Iraq run by a unit known as Brigade 30. It housed about one thousand detainees arrested on sectarian charges. See "Horrific Testimonies, Secret Prisons Portend Catastrophe in Iraq," *Euro-Med Monitor*, July 13, 2020, https://reliefweb.int/report/iraq/horrific-testimonies-secret-prisons-portend-catastrophe-iraq. Other secret prisons were reportedly under the control of the prime minister's office. See "Iraq: Secret Jail Uncovered in Baghdad," *Human Rights Watch*, February 1, 2011, https://www.hrw.org/news/2011/02/01/iraq-secret-jail-uncovered-baghdad#. See also Florence Gaub, "An Unhappy Marriage: Civil-Military Relations in Post-Saddam Iraq," *Regional Insight*, Carnegie Europe, January 13, 2016, https://carnegieeurope.eu/2016/01/13/unhappy-marriage-civil-military-relations-in-post-saddam-iraq-pub-61955.

21. For example, see "Iraq Army 'Had 50,000 Ghost Troops' on Payroll," BBC.com, November 30, 2014, https://www.bbc.com/news/world-middle-east-30269343.

22. Querine Hanlon, "Security Sector Reform in Tunisia: A Year after the Jasmine Revolution," Special Report no. 304, United States Institute of Peace, March 2012.

23. For a detailed plan, see Strategic Capacity Group (SCG), *Planning for Disarmament, Demobilization, and Reintegration (DDR) and Security Sector Reform (SSR) in Libya: A Building Block Approach* (Washington, DC: SCG, September 2020).

24. SCG, *Abbreviated Capacity Assessment of the Government of National Accord's Security Ministries* (Washington, DC: SCG, July 2020), 3.

25. Ibid., 5.

26. Ibid., 1.

27. European External Action Service, "EUBAM Libya Initial Mapping Report," EUBAM Libya (Brussels: European External Action Service, 2017); and SCG interview with EUBAM officials, Tunis, March 8, 2018.

28. Querine Hanlon and Alexandra Kerr, *North Africa Regional Border Security Assessment* (McLean, VA: Strategic Capacity Group, 2019), 12.

29. Interview with Maka Petriashvili, Georgia Ministry of Defense official and SCG researcher, Tblisi, April 29, 2017. See also David Darchiashvili, "Georgian Security Sector: Achievements and Failures," in *Security Sector Governance in Southern Caucasus: Challenges and Visions*, ed. Anja H. Ebnöther and Gustav E. Gustenau (Vienna: National Defence Academy and Bureau for Security Policy, in cooperation with the PfP Consortium of Defence Academies and Security Studies Institutes: Study Group Information, 2004), 84–114.

30. Petriashvili, interview.

31. Ibid.

32. Interview with Teona Akubardia, deputy secretary, National Security Council of Georgia, Tblisi, March 18, 2017.

33. Mindia Vashakmadze, *The Legal Framework of Security Sector Governance in Georgia*, Geneva Centre for Security Sector Governance (DCAF), 2014, https://www.dcaf.ch/legal-framework-security-sector -governance-georgia.

34. Born, "Democratic Control of Defence Activities," 78–79.

35. Transparency International, "Transparency International Defense and Security, Colombia: Government Defense and Anti-Corruption Index, 2016," December 28, 2016, http://ti-defence.org/wp-content/uploads /2016/09/GI-Colombia-2016.pdf.

36. David Darchiashvili, "Georgian Defense Policy and Military Reform," in *Statehood and Security: Georgia after the Rose Revolution*, ed. B. Coppieters and R. Legvold (Boston: MIT Press, 2005), 124–125.

37. Ibid.

38. Atlantic Council of Georgia, *Review of Georgia's National Security Architecture: Strategic Level*, Security Sector Review Project, November 2013, http://acge.ge/2015/04/review-of-georgias-national-security -architecture-strategic-level-2/.

Chapter 4. Goal 2: Civilian Control

1. Fred Schreier, "The Division of Labour in the Defence and Security Sphere," in *Defence Institution Building: A Sourcebook in Support of the Partnership Action Plan (PAP-DIB)*, ed. Wim F. Van Eekelen and Philipp H. Fluri, 19–77 (Vienna: Austrian National Defence Academy; Geneva: Geneva Centre for the Democratic Control of Armed Forces, 2006), 26

2. Ibid., 27.

3. Susan Pond, "Partnership Action Plan on Defense Institution Building: Concept and Implementation," in *Defence Institution Building: 2005 Partnership Action Plan on Defence Institution Building Regional Conference*, ed. Philipp Fluri and Eden Cole, 4–13 (Vienna: Austrian National Defence Academy; Geneva: Geneva Centre for the Democratic Control of Armed Forces Defence, 2005), 4.

4. Samuel Huntington, *The Soldier and the State: The Theory and Politics of Civil-Military Relations* (Cambridge, MA: Belknap Press of Harvard University Press, 1957), 2.

5. Ibid..

6. Willem F. Van Eekelen, "Civil-Military Relations and the Formulation of Security Policy," in *Defence Institution Building: A Sourcebook in Support of the Partnership Action Plan (PAP-DIB)*, ed. Wim F. Van Eekelen and Philipp H. Fluri, 108–137 (Vienna: Austrian National Defence Academy; Geneva: Geneva Centre for the Democratic Control of Armed Forces, 2006), 109.

7. Ibid.

8. Ibid., 109–110.

9. Ibid.,111.

10. Zoltan Barany, *The Soldier and the Changing State: Building Democratic Armies in Africa, Asia, Europe, and the Americas* (Princeton: Princeton University Press, 2012), 1.

11. Van Eekelen, "Civil-Military Relations," 113.

12. Ibid.

13. Ibid.

14. Ibid., 129.

15. Ibid.,129–130.

16. Government of Georgia, The Regulation of the Ministry of Defense, 2013. See also Ministry of Defense of Georgia, "Defense Forces," https://mod.gov.ge/ge/page/13/tavdacvis-dzalebi.

17. Parliament of Georgia, *Law on Defense of Georgia*, 1997, Chapter 3, Article 7, https://matsne.gov.ge/ka /document/view/28330.

18. Although Tunisian government websites such as www.defense.tn may not feature robust information about the defense sector, other publicly available sources do. For example, the Geneva Centre for Security Sector Governance (DCAF) maintains a website listing all legislative and regulatory texts governing Tunisia's security sector. See https://legislation-securite.tn/.

19. Hanlon, "Security Sector Reform in Tunisia," 4.

20. Ibid., 5.

21. Interview with Shalva Dzebisashvili, director of the Defense Policy and Development Department, Ministry of Defense of Georgia, Tblisi, March 6, 2017.

22. Interview with Ministry of Defense officers, Tunis, 2016.

23. Interview with a military officer, Tunis, 2016.

24. For the text of the 2005 constitution, see Constituteproject.org, last modified April 9, 2021, https://www .constituteproject.org/constitution/Iraq_2005.pdf?lang=en.

25. Gaub, "An Unhappy Marriage."

26. Ibid.

27. Ibid.

28. Ibid.

29. Ibid. See also Abeer Mohammed and Katherine Zoepf, "Iraq's Defense Minister Commits Employees to Political Neutrality," *New York Times*, October 31, 2008, http://www.nytimes.com/2008/10/31/world /africa/31iht-iraq.1.17414956.html.

30. Gaub, "An Unhappy Marriage."

31. Interview with David Gunashvili, director of the Human Resource Management and Professional Development Department, Ministry of Defense of Georgia, March 11, 2017.

32. Interviews with defense sector officials, conducted by Olga Nazario, Bogota, Colombia, May 2017.

33. Ibid.

34. Interview with Nodar Kharshiladze, head of the Strategic Analysis Center and former deputy minister of defense of Georgia, Tblisi, March 8, 2017; interview with Teona Akubardia, deputy chair of the National Security Council of Georgia, Tblisi, March 18, 2017; and interview with Dzebisashvili, March 6, 2017.

35. Transparency International, "Transparency International Defense and Security, Colombia."

36. Human Rights Watch, "Colombia: Events of 2016," *Human Rights Watch World Report 2017*, https://www .hrw.org/world-report/2017/country-chapters/colombia.

37. "Linea del Honor-163," Policia.gov., accessed August 31, 2020, https://www.policia.gov.co/noticia/l%C3 %ADnea-del-honor%E2%80%93-163.

38. SCG interview with defense sector reform expert, Tunis, March 19, 2017.

39. Ibid.

40. Ibid.

41. Interview with the director of the Tunisian Center for Global Research, Tunis, March 18, 2017.

42. Interview with the director of ITES, Tunis, March 19, 2017.

Chapter 5. Goal 3: Legislative and Judicial Oversight

1. Born, "Democratic Control of Defence Activities," 78–79.

2. Ibid.

3. Decree 1050 (2015); Law 1407 (2010), §§ 1 et seq.

4. A widely reported case from 2008 involved soldiers and officers who abducted victims or lured them to remote locations under false pretenses—such as promises of work—and killed them, placed weapons on their bodies, and reported them as enemy combatants killed in action. See Human Rights Watch, "Colombia: Events of 2016," *Human Rights Watch World Report 2017*, https://www.hrw.org/world-report/2017/country -chapters/colombia.

5. D. Valero, "Quedo listo revolcon de la justiciar penal milititar" [The Military Criminal Justice Revolt Was Ready], *El Tiempo*, accessed July 27, 2015, http://www.eltiempo.com/archivo/documento/CMS-16153367.

6. "Santos asegura que militares necesitan proteccion juridical fuerte" [Santos Assures that the Military Needs Strong Legal Protection], *El Tiempo*, accessed September 9, 2014, http://www.eltiempo.com/archivo/ documento/CMS-14505041.

7. Interview with José Luis Pérez Oyuela, Bogotá, May 17, 2017.

8. Constitution of Colombia, Articles 114, 138, 174, 175, 178, and 235.

9. Transparency International, "Transparency International Defense and Security, Colombia."

10. David Darchiashvili, "Georgia: A Hostage to Arms," in *The Caucasus: Armed and Divided—Small Arms and Light Weapons Proliferation and Humanitarian Consequences in the Caucasus*, ed. Duncan Hiscock and Anna Matveeva, 1–33 (London: Saferworld, 2003),17.

11. Some estimates suggest that only 40 percent of military officers on the payroll actually reported for duty. Petriashvili interview.

12. Darchiashvili, "Georgia: A Hostage to Arms," 13.

13. Ibid.

14. Ibid.

15. Ibid., 18.

16. Ibid.

17. Mindia Vashakmadze, *The Legal Framework of Security Sector Governance in Georgia* (Geneva: Geneva Centre for the Democratic Control of Armed Forces, 2014), 9.

18. Ibid., 11.

19. Interview with Sopo Babunashvili, head of Apparatus of the Parliament Defense and Security Committee, Tblisi, March 13, 2017.

20. Interview with David Darchiashvili, head of the Civil-Military Relations Center and former Member of Parliament, March 15, 2017; and interview with Bruce Bach, expert, US Defense Institution Building Advisory Team, March 14, 2017.

21. Interviews with Akubardia; William Lahue, head of NATO Liaison Office in Georgia, March 9, 2017; Alex Lewis Paul, UK Special Defence Advisor to the Minister of Defense of Georgia, March 15, 2017; Bruce Bach, expert, US Defense Institution Building Advisory Team, March 14, 2017; Giorgi Shaishmelashvili, deputy director of the Defense Policy and Development Department, Ministry of Defense of Georgia, March 6, 2017; Major General Vakhtang Kapanadze, senior military representative of Georgia to NATO and former chief of the General Staff of the Georgian Armed Forces, March 18, 2017; and Giorgi Muchaidze, Executive Director of Atlantic Council, Georgia, former Minister of Defense of Georgia, March 14, 2017.

22. U.S. Department of Defense, "Measuring Stability and Security in Iraq," Report to the U.S. Congress in accordance with the Department of Defense Supplemental Appropriations Act 2008 (Section 9204, Public Law 110-252), March 2010, 46.

23. Interview with Ajmi Lourimi, member of parliament, Tunis, March 17, 2017.

24. Ibid.

25. Ibid.

26. Panel discussion with members of the Ad Hoc Committee on Administrative Reform and Good Governance, Tunis, March 14, 2017.

27. Interview with a member of the Security and Defense Committee, Tunis, March 17, 2017.

28. Ibid.

29. Lourimi, interview.

30. Ibid.

31. Ibid.

32. Panel discussion with members of the Ad Hoc Committee on Administrative Reform and Good Governance, Tunis, Tunisia, March 14, 2017.

33. Ibid.

34. Interview with members of the Security and Defense Committee, Tunis, March 14–17, 2017.

35. Ibid.

36. Interview with Ousama Al Saghir, member of the Security and Defense Committee, Tunis, March 17, 2017.

37. Interview with Mokhtar ben Nasr, director, Tunisian Center for Global Research, Tunis, March 18, 2017.

38. Interview with Tina Khidasheli, former minister of defense, Tblisi, March 14, 2017.

39. Muchaidze, interview.

40. Interview with Tamar Karosanidze, chief of party of East-West Management Institute and former deputy minister of defense, Tblisi, Georgia, March 10, 2017.

41. Babunashvili, interview.

42. Akubardia, interview.

43. Interview with Vepkhvia Grigalashvili, deputy director of the legal department of the Ministry of Defense, Tblisi, March 9, 2017; and Babunashvili, interview.

Chapter 6. Goal 4: Coordination and Management

1. Tom Galvin et al. eds., *Defense Management: Primer for Senior Leaders* (Carlisle, PA: United States Army War College, May 26, 2018), https://publications.armywarcollege.edu/pubs/3534.pdf.

2. Philipp Fluri and Eden Cole, eds., *2005 Partnership Action Plan on Defence Institution Building Regional Conference* (Vienna: Austrian National Defence Academy; Geneva: Geneva Centre for the Democratic Control of Armed Forces Defence, 2005), 8.

3. Hari Bucur-Marcu, ed., *Essentials of Defence Institution Building* (Vienna: Austrian National Defence Academy; Geneva: Geneva Centre for the Democratic Control of Armed Forces Defence, May 22, 2009), 75.

4. Ibid., 79.

5. Ibid., 76.

6. Ibid., 77.

7. For example, civilians competing for senior management positions in the U.S. Department of Defense must undergo requisite management and coordination training to join the Senior Executive Service. See U.S. Office of Personnel Management, "Senior Executive Service: Candidate Development Programs," https://www.opm.gov/policy-data-oversight/senior-executive-service/candidate-development-programs.

8. Schreier, "The Division of Labour in the Defense and Security Sphere," 24.

9. Parliament of Georgia, "Law of Georgia on Planning and Coordination of the National Security Policy," *Legislative Herald of Georgia*, March 2015, https://matsne.gov.ge/en/document/view/2764463. See also Teona Akubardia, "National Security Policy: Planning, Coordination and Practice in Georgia," Expert Opinion 110 (Tbilisi: Georgian Foundation for Strategic and International Studies, 2018).

10. Ibid.

11. Petriashvili, interview.

12. Akubardia, "National Security Policy," 5.

13. Interview with government ministry personnel, Bamako, September 12–20, 2016, and October 20, 2016.

14. Office of Security Cooperation–Iraq, "Five Year Security Cooperation and Assistance Roadmap for Iraq," Baghdad, February 15, 2018 (unpublished report); and Renad Mansour and Erwin Van Veen, "Iraq's Com-

peting Security Forces after the Battle for Mosul," *War On the Rocks*, last modified August 25, 2017, https://warontherocks.com/2017/08/iraqs-competing-security-forces-after-the-battle-for-mosul/.

15. Office of Security Cooperation-Iraq, "Five Year Security Cooperation and Assistance Roadmap for Iraq."

16. "Section 3: National Security Adviser," Coalition Provisional Authority Order no. 68, https://nsa.gov.iq/page/2/incorporation-law.

17. Interview with U.S. Security Cooperation official in Iraq, January 29, 2018.

18. Government of Colombia, Law 489, Article 59 (1998) and Decree 1512 (2000).

19. Ibid.

20. For a defense sector reform plan, see Strategic Capacity Group, *Planning for Disarmament, Demobilization, and Reintegration (DDR) and Security Sector Reform (SSR) in Libya: A Building Block Approach* (Washington, DC: Strategic Capacity Group, September 2020).

21. Strategic Capacity Group (SCG), *Abbreviated Capacity Assessment of the Government of National Accord's Security Ministries* (Washington, DC: Strategic Capacity Group, July 2020), 3.

22. Ibid., 5

23. Darchiashvili, "Georgia: A Hostage to Arms," 13.

24. Ibid.

25. Interview with U.S. Security Cooperation official in Iraq, January 29, 2018. See also C. Anthony Pfaff, *Professionalizing the Iraqi Army: US Engagement after the Islamic State* (Carlisle, PA: United States Army War College Press, January 2020), 91.

Chapter 7. Goal 5: Functioning Logistics

1. Martin Van Creveld, *Supplying War: Logistics from Wallenstein to Patton*, (Cambridge: Cambridge University Press, 1977), 1.

2. NATO, "Troop Contributions," North Atlantic Treaty Organization, March 23, 2020, last modified March 23, 2020, https://www.nato.int/cps/en/natohq/topics_50316.htm.

3. U.S. Army Maneuver Center of Excellence, "Military Logistics," Maneuver Self Study Program, last modified December 18, 2018, https://www.benning.army.mil/mssp/Logistics/#:~:text=Transportation%20includes%20the%20movement%20of,subsequently%20maintains%20and%20secures%20them.

4. Ibid.

5. Michael Boomer and George Topic, "Logistics," in *Effective, Legitimate, Secure: Insights for Defense Institution Building*, ed. Alexandra Kerr and Michael Miklaucic, 139–159 (Washington, DC: National Defense University, 2018), 150.

6. Boomer and Topic, "Logistics," 144–145.

7. John Ismay, "The U.S. Spends Billions in Defense Aid: Is It Working?," *New York Times*, June 13, 2018, https://www.nytimes.com/2018/06/13/magazine/train-equip-defense-aid.html.

8. UN, Department of Field Support, *Defence Sector Reform*, 5.

9. Office of the Secretary of Defense for Policy, "DOD Directive 5205.82 Defense Institution Building (DIB)," 4.

10. Willem F. Van Eekelen, "Parliaments and Defence Spending," in *Defence Institution Building: A Sourcebook in Support of the Partnership Action Plan (PAP-DIB)*, ed. Willem F. Van Eekelen and Philipp H. Fluri, 365–396 (Vienna: Austrian National Defence Academy; Geneva: Geneva Centre for the Democratic Control of Armed Forces, 2006), 391.

11. Mihaly Zambori, "Economically Viable Management and Defence Spending," in *Defence Institution Building: A Sourcebook*, 275–293, 291–292.

12. Eekelen, "Parliaments and Defence Spending," 370.

13. U.S. Army Maneuver Center of Excellence, "Military Logistics."

14. FORSCOM for STAND-TO!, "Army Force Generation," www.army.mil, last modified July 29, 2010, https://www.army.mil/article/42519/army_force_generation.

15. Robert Curley, ed., *The Science of War: Strategies, Tactics, and Logistics* (Chicago: Britannica Educational Publishing, 2011), 71.

16. U.S. Department of the Army, *Army Doctrine Publication No. 4-0 Sustainment* (Washington, DC: U.S. Department of the Army Headquarters, 2019), 7–8.

17. Ibid.,1–5.

18. NATO Logistics Committee, *NATO Logistics Handbook* (Brussels: NATO Headquarters, 2012), 115.

19. Office of the Deputy Assistant Secretary of Defense for Security Cooperation, *Defense Institution Building Handbook: A Practical Guide* (Washington, DC: Office of the Undersecretary of Defense for Policy, 2017), 7–9.

20. Robbin Laird and Ed Timperlake, "The French in Mali: Shaping the Logistics Element of the Operation," *SLDinfo.com*, last modified May 6, 2013, https://sldinfo.com/2013/05/the-french-in-mali-shaping-the-logistics-element-of-the-operation/.

21. Thomas W. Ross, Jr., "Defining the Discipline in Theory and Practice," in *Effective, Legitimate, Secure: Insights for Defense Institution Building*, ed. Alexandra Kerr and Michael Miklaucic, 21–46 (Washington, DC: National Defense University, 2015), 21.

22. Chivers, "After Retreat, Iraqi Soldiers Fault Officers."

23. Boomer and Topic, "Logistics," 150.

24. Ross, Jr., "Defining the Discipline in Theory and Practice," 21.

25. U.S. Department of State, "U.S. Security Cooperation with Iraq," United States Department of State, Bureau of Political-Military Affairs, last modified January 20, 2021, https://www.state.gov/u-s-security-cooperation-with-iraq/.

26. Ibid. By July 2021, the total amount had changed to $16.3 billion.

27. See, for example: Eric Peltz et. al., *Sustainment of Army Forces in Operation Iraqi Freedom: Major Recommendations* (Santa Monica, CA: RAND Corporation, 2019), 4; and William M. Solis, *Defense Logistics: Preliminary Observations on the Effectiveness of Logistics Activities During Operation Iraqi Freedom* (Washington, DC: U.S. Government Accountability Office, 2018).

28. Interview with Michael D. Sullivan, U.S. Army colonel stationed at the Office of Security Cooperation–Iraq in 2018, in McLean, Virginia, February 18, 2021.

29. Cliff Hinote and Michael Sullivan, *The Five-Year Security Cooperation and Assistance Roadmap for Iraq* (Baghdad: Office of Security Cooperation–Iraq, 2018).

Chapter 8. Goal 6: Defense Planning

1. Paul K. Davis, "Defense Planning When Major Changes Are Needed," *Defence Studies* 18 (2018), 374–390.

2. Geneva Centre for Security Sector Governance (DCAF), Building Integrity Programme, "Defense Planning," Security Sector Integrity, accessed February 22, 2021, https://securitysectorintegrity.com/defence-management/planning/.

3. Note that a third category, tactical planning, is sometimes differentiated from these two, though more often it is considered a subset of operational planning.

4. Mihaly Zambori, "Economically Viable Management and Defence Spending," in *Defence Institution Building: A Sourcebook in Support of the Partnership Action Plan (PAP-DIB)*, ed. Willem F. Van Eekelen and Philipp H. Fluri, 275–291 (Vienna: Austrian National Defence Academy; Geneva: Geneva Centre for the Democratic Control of Armed Forces, 2006), 284.

5. Zambori, "Economically Viable Management and Defence Spending," 284.

6. Jans Arveds Trapans, "Democracy, Security, and Defence Planning," in *Defence Institution Building: A Sourcebook in Support of the Partnership Action Plan (PAP-DIB)*, ed. Willem F. Van Eekelen and Philipp H. Fluri, 184–205 (Vienna: Austrian National Defence Academy; Geneva; Geneva Centre for the Democratic Control of Armed Forces, 2006), 197.

7. See chapter 11 for additional discussion on the meaning and role of defense strategy in the defense sector.

8. Michael J. Mazarr et al., *The U.S. Department of Defense's Planning Process: Components and Challenges* (Santa Monica, CA: RAND Corporation, 2019).

9. Trapans, "Democracy, Security, and Defence Planning," 192.

10. Zambori, "Economically Viable Management and Defence Spending," 284.

11. Ibid., 278–279.

12. DCAF Building Integrity Programme, "Defense Planning."

13. OECD, *OECD DAC Handbook on Security System Reform*, 67; and Office of the Secretary of Defense for Policy, "DOD Directive 5205.82 Defense Institution Building (DIB)," 4.

14. Philipp H. Fluri, "Preface," in *Defense Institution Building: 2005 Partnership Action Plan on Defence Institution Building Regional Conference*, ed. Philipp H. Fluri and Eden Cole, 1–2 (Vienna: Austrian National Defence Academy; Geneva: Geneva Centre for the Democratic Control of Armed Forces), 2.

15. Susan Pond, "Partnership Action Plan on Defence Institution Building: Concept and Implementation," in *Partnership Action Plan on Defence Institution Building: Concept and Implementation*, ed. Philipp H. Fluri and Eden Cole, 4–13 (Vienna: Austrian National Defence Academy; Geneva: Geneva Centre for the Democratic Control of Armed Forces, 2005), 11.

16. Mark Gilchrist, "It's a Journey, Not a Destination: Seven Lessons for Military Planners," Modern War Institute at West Point, last modified September 17, 2019, https://mwi.usma.edu/journey-not-destination-seven-lessons-military-planners/.

17. Zambori, "Economically Viable Management and Defence Spending," 278–279.

18. NATO, "NATO Defence Planning Process," last modified May 11, 2021, https://www.nato.int/cps/en/natohq/topics_49202.htm.

19. Hari Bucur-Marcu, "Financial Planning and Resource Allocation in the Defence Area," in *Defence Institution Building: A Sourcebook in Support of the Partnership Action Plan (PAP-DIB)*, ed. Willem F. Van Eekelen et al., 255–274 (Vienna: Austrian National Defence Academy; Geneva: Centre for the Democratic Control of Armed Forces, 2006), 258.

20. DCAF Building Integrity Programme, "Defense Planning."

21. Zambori, "Economically Viable Management and Defence Spending," 278–279.

22. Ibid.

23. Ibid.

24. United Nations, "Lesson 1: Military Planning Process," 15–17, 23–26, 73, http://dag.un.org/bitstream/handle/11176/400722/M3%20Operational.pdf?sequence=4&isAllowed=y.

25. Trapans, "Democracy Security, and Defence Planning," 197.

26. DCAF Building Integrity Programme, "Defense Planning."

27. Joseph L. Derdzinski, "Executive Summary: PAP-DIB Sourcebook," in *Defence Institution Building: A Sourcebook*, 455–510, 489.

28. Trapans, "Democracy Security, and Defence Planning," 197.

29. DCAF Building Integrity Programme, "Defense Planning."

30. Trapans, "Democracy Security, and Defence Planning," 197.

31. Ibid.

32. Todor Tagarev, "Defence Planning: Core Processes in Defence Management," in *Defence Management: An Introduction*, ed. Hari Bucur-Marcu and Todor Tagarev, 45–74 (Geneva: Geneva Center for the Democratic Control of Armed Forces, 2009), 45.

33. Parliament of Georgia, Law of Georgia on Defense Planning, Pub. L. No. 2956, §9, first published 2006; consolidated versions published September 27, 2013–October 31, 2018, https://matsne.gov.ge/en/document/view/26230?publication=2.

34. Ibid., Art. 5, §9.

35. Ibid., Art. 10.

36. Ibid.

37. Ibid.

38. Interview with Giorgi Dolidze, head of the Planning, Programming, and Budgeting Division in the Defense Policy and Development Department of the Ministry of Defense of Georgia, Tbilisi, March 6, 2017.

39. Trapans, "Democracy Security, and Defence Planning," 191.

40. Interview with Vasil Garsevanishvili, deputy chief of the J1 Personnel Management Department of the General Staff, Georgian Armed Forces, Tbilisi, March 7, 2017.

41. Interview with Colonel Zviad Shanava, chief of J4 Logistics Department of the General Staff, Georgian Armed Forces, Tbilisi, March 9, 2017.

42. Michael D. Sullivan, "The Methodological Assumption of Baghdad," in *Shoulder to Shoulder: Achieving Peace through Partnership*, ed. Gary J. Morea, 56–87 (unpublished draft, last modified 2010, copy in author's files), 60.

43. U.S. Army Combined Arms Center, "Converting Intellectual Power into Combat Power," School of Advanced Military Studies, last modified April 25, 2019, https://usacac.army.mil/organizations/cace/cgsc/sams.

44. U.S. Department of State, "U.S. Security Cooperation with Iraq," United States Bureau of Political-Military Affairs Fact Sheet, last modified January 20, 2019, https://www.state.gov/u-s-security-cooperation-with-iraq-2/.

45. Ministry of Defense of Georgia, "Structure of Defense of Georgia" 2017, https://mod.gov.ge/uploads/2019/Structure/structure_en_04_08_2020.pdf.

Chapter 9. Goal 7: Financial Management

1. See, for example, Congressional Research Services (CRS), *Defense Primer: Planning, Programming, Budgeting, and Execution (PPBE) Process* (Washington, DC: CRS, January 27, 2020), https://fas.org/sgp/crs/natsec/IF10429.pdf.

2. Willem F. Van Eekelen, "Parliaments and Defense Procurement," in *Defence Institution Building: A Sourcebook, in Support of the Partnership Action Plan (PAP-DIB)* ed. Wim F. Van Eekelen and Philipp H. Fluri, 420–433 (Vienna: National Defence Academy; Geneva, Centre for Democratic Control of Armed Forces, 2006), 422–424.

3. For example, see chapters 9, 10, and 13 in Wim F. Van Eekelen and Philipp H. Fluri, eds., *Defence Institution Building: A Sourcebook*; and chapters 8 and 9 in Hari Bucur-Marcu, ed., *Essentials of Defence Institution Building* (Vienna: National Defence Academy, 2009).

4. OECD, *OECD DAC Handbook on Security System Reform*, 93.

5. Hari Bucur-Marcu, "Financial Planning within Defense," in *Essentials of Defence Institution Building*, ed. Hari Bucur-Marcu, 103–113 (Vienna: National Defence Academy, 2009), 108.

6. Ibid.

7. Ibid., 108–109.

8. Ibid., 112.

9. Ibid., 110.

10. Hari Bucur-Marcu, "Financial Planning and Resource Allocation in the Defence Area," in *Defence Institution Building: A Sourcebook in Support of the Partnership Action Plan (PAP-DIB)*, ed. Wim F. Van Eekelen and Philipp H. Fluri, 255–274 (Vienna: National Defence Academy, 2006), 256.

11. Ibid., 262.

12. Ibid.

13. Interview with David Gunashvili, director of the Human Resource Management and Professional Development Department, Ministry of Defense of Georgia, March 11, 2017.

14. Interview with Giorgi Dolidze, head of the Planning, Programming, and Budgeting Division in the Defense Policy and Development Department of the Ministry of Defense of Georgia, Tbilisi, March 11, 2017.

15. Interview with Colonel Nick Janjghava, first deputy of the Chief of General Staff of the Georgian Armed Forces, Tbilisi, March 17, 2017.

16. Interview with Martha Rueda, adviser at the Finance Department of the National Army, Bogotá, May 18, 2017.

17. Ibid.

18. United States Agency for International Development (USAID) Colombia, "Colombia Program At-a-Glance," 2014, https://www.usaid.gov/sites/default/files/Colombia%20Country%20Fact%20Sheet%20Augst%202013_USAID_at_a_Glance.pdf.

19. USAID, "U.S. Foreign Aid by Country: Colombia," Foreign Aid Explorer, last modified February 25, 2021, https://explorer.usaid.gov/cd/COL?fiscal_year=2017&measure=Obligations.

20. Army of Colombia, *Manual Fundamental de Operaciones 3.0* (Bogota: Army of Colombia, 2016), 12.

21. Transparency International Defence and Security, *Government Defence Anti-Corruption Index: Colombia* (Bogota: Transparency International, 2016), 37.

22. Ibid., 30.

23. Ibid., 2.

24. Developed by the European Commission, the PIFC concept provides structural and operational models to help national governments upgrade their public control systems to meet EU and international standards.

25. See State Procurement Agency of Georgia, "Functions," 2016, http://www.procurement.gov.ge/ka/page/features; and State Procurement Agency of Georgia, "Structure," 2016, http://www.procurement.gov.ge/ka/page/aboutagency.

26. The department submits tender announcements to the State Procurement Agency and the Georgian Unified Electronic Government Procurement System.

27. State Procurement Agency of Georgia, *Unified Electronic System of State Procurement: User Manual* (Tbilisi: State Procurement Agency of Georgia, 2015); and State Procurement Agency of Georgia, "Civil-Society Cooperation," http://procurement.gov.ge/Civil-Society-Cooperation.aspx.

28. Interview with Nino Karazanashvili, head of the Internal Audit Department of the Ministry of Defense of Georgia, Tbilisi, March 22, 2017.

29. Ibid.

30. "Libya's U.N.-Backed Government Steps Up Defense Spending as War Drags On," Reuters, last modified August 6, 2019, https://www.reuters.com/article/us-libya-security/libyas-u-n-backed-government-steps-up-defense-spending-as-war-drags-on-idUSKCN1UW1Z4.

31. Transparency International Defence and Security, *Government Defence Anti-Corruption Index-Colombia*, 22.

32. Ibid.

33. Rueda, interview.

Chapter 10. Goal 8: The Right People

1. James Greene, "Personnel Policies," in *Building Integrity and Reducing Corruption in Defence: A Compendium of Best Practices*, ed. Todor Tagarev, 43–56, (Geneva: Geneva Centre for the Democratic Control of Armed Forces, 2010), 43.

2. Office of the Secretary of Defense for Policy, "DOD Directive 5205.82 Defense Institution Building (DIB)," 4.

3. Philipp H. Fluri, "Preface," in Philipp H. Fluri and Eden Cole, eds., *Defence Institution Building: 2005 Partnership Action Plan on Defence Institution Building Regional Conference* (Vienna: Ministry of Defence; Geneva, Centre for Democratic Control of Armed Forces, 2005), 1.

4. OECD, *OECD DAC Handbook on Security System Reform*, 14, 65, 94, 102.

5. Ferenc Molnar, "Principles and Practices in Personnel Policies: The Case of the Hungarian Defence Forces," in *Defence Institution Building: A Sourcebook in Support of the Partnership Action Plan (PAP-DIB)*, ed.

Willem F. Van Eekelen et al., 234–254 (Vienna: Ministry of Defence; Geneva, Centre for Democratic Control of Armed Forces, 239–240.

6. OECD, *OECD DAC Handbook on Security System Reform*, 66, 94.

7. UN, Department of Field Support, *Defence Sector Reform*, 23.

8. Susan Pond, "Partnership Action Plan on Defence Institution Building: Concept and Implementation," in *Partnership Action Plan on Defence Institution Building: Concept and Implementation*, ed. Philipp H. Fluri et al. (Vienna: Ministry of Defence; Geneva: Geneva Centre for the Democratic Control of Armed Forces, 2005), 9.

9. Simon Lunn, "Parliamentary and Executive Oversight of the Defence Sphere," in *Defense Institution Building: 2005 Partnership Action Plan on Defence Institution Building Regional Conference*, ed. Philipp H. Fluri et al., 32–48 (Vienna: National Defence Academy; Geneva: Centre for the Democratic Control of Armed Forces, 2005), 34.

10. Molnar, "Principles and Practices in Personnel Policies," 235.

11. Lunn, "Parliamentary and Executive Oversight of the Defence Sphere," 45–46.

12. OECD, *OECD DAC Handbook on Security System Reform*, 173.

13. Ibid., 168.

14. Ibid., 109–110, 167.

15. Ibid., 176.

16. UN, Department of Field Support, *Policy: Defence Sector Reform*, 10.

17. Molnar, "Principles and Practices in Personnel Policies," 244.

18. Pond, "Partnership Action Plan on Defence Institution Building," 9.

19. OECD, *OECD DAC Handbook on Security System Reform*, 167.

20. Andrzej Karkoszka, "Transparency and Accountability in Defence Management, in *Defense Institution Building: 2005 Partnership Action Plan on Defence Institution Building Regional Conference*, ed. Philipp H. Fluri et al., 56–61 (Vienna and Geneva: National Defence Academy Geneva and Geneva Centre for the Democratic Control of Armed Forces, 2005), 55–56.

21. Congressional Research Service (CRS) (Kristy N. Kamarck, author), *Military Retirement: Background and Recent Developments* (Washington, DC: CRS, 2019), 1, https://fas.org/sgp/crs/misc/RL34751.pdf.

22. OECD, *OECD DAC Handbook on Security System Reform*, 132.

23. CRS, "Military Retirement:," 1.

24. OECD, *OECD DAC Handbook on Security System Reform*, 94.

25. For additional information on developing strategic thinking, see chapter 11 in this volume, "Strategy Generation."

26. Querine Hanlon and Alexandra Kerr, *North Africa Regional Border Security Assessment* (McLean, VA: Strategic Capacity Group, 2019), 41.

27. Parliament of Georgia, "Types and Composition of the Defence Forces of Georgia," "Law of Georgia on the Defense of Georgia," Ch. III, Art. 9, first published 1997, consolidated version July 17, 2020.

28. Janjghava, interview. For the Strategic Defense Review and Implementation Plan for 2017–20, see https://mod.gov.ge/uploads/2018/pdf/SDR-ENG.pdf; for the Georgian National Military Strategy for 2014, see https://mod.gov.ge/uploads/2018/pdf/NMS-ENG.pdf; and for the Threat Assessments for 2010–13, see https://mod.gov.ge/uploads/2018/pdf/TAD-ENG.pdf.

29. Garsevanishvili, interview.

30. Janjghava, interview.

31. Hanlon and Kerr, *North Africa Regional Border Security Assessment*, 41.

32. Christie Caldwell, "Key Talent Considerations for Working in Iraq," SHRM, last modified April 19, 2013, https://www.shrm.org/resourcesandtools/hr-topics/global-hr/pages/talent-management-iraq.aspx.

33. United Nations Office on Drugs and Crime (UNODC), *Corruption and Integrity Challenges in the Public Sector of Iraq* (New York: UNODC, 2013), 12, 37.

34. Ibid., 56.

35. Interview with Adam Thian, journalist, in Bamako, September 14, 2016; and SCG interview with Kissima Gakou, former technical adviser to the Malian Ministry of Defense, Bamako, September 14, 2016.

36. Interview with Shalva Dzebisashvili, director of the Defense Policy and Development Department, Ministry of Defense of Georgia, Tbilisi, March 6, 2017.

37. Adriana Lins de Albuquerque and Jakob Hedenskog, *Georgia: A Defence Sector Reform Assessment* (Stockholm: Swedish Defence Research Agency, 2016), 34.

38. Ibid.

39. Ibid.

40. Ibid., 32.

41. Ibid., 33.

42. Ibid.

43. Interview with Martha Rueda, adviser at the Finance Department of the National Army of Colombia, in Bogota, Colombia, May 15, 2017.

44. Hanlon and Kerr, *North Africa Regional Border Security Assessment*, 41.

45. Ibid.

46. Parliament of Georgia: "Law of Georgia on the Defence of Georgia," Pub. L. No. 1030, Ch. III, Art. 9.

47. Ibid.

48. Ministry of Defense of Georgia, "Vision of the Training and Military Education Command," https://mod.gov.ge/en/page/57/vision-of-the-training-and-military-education-command.

49. Interview with Colonel Malkhaz Makaradze, commander of the Education and Military Training Command, Georgian Armed Forces, Tbilisi, March 17, 2017.

50. Ibid.

51. Ibid.

Chapter 11. Goal 9: Strategy Generation

1. Office of the Secretary of Defense for Policy, "DOD Directive 5205.82 Defense Institution Building (DIB)," 4; and UN, Department of Field Support. *Defence Sector Reform*, 6.

2. OECD, *OECD DAC Handbook on Security System Reform*, 92.

3. UN, Department of Field Support, *Defence Sector Reform*, 6.

4. NATO, *NATO Partnership Action Plan on Defense Institution Building*, 2.

5. NATO, *NATO Partnership Action Plan on Defense Institution Building Reference Curriculum*, (Ontario: Partnership for Peace Consortium of Defense Academies and Security Studies Institutes, 2008), 23.

6. Heather Wolters, Anna Grome, and Ryan Hinds, eds., *Exploring Strategic Thinking: Insights to Assess, Develop, and Retain Army Strategic Thinkers* (Fort Belvoir, VA: United States Army Research Institute for the Behavioral and Social Sciences, 2013), 248.

7. Joint Chiefs of Staff, "Strategy," Joint Doctrine Note 2-19, December 10, 2019, I-3, https://www.jcs.mil/Portals/36/Documents/Doctrine/jdn_jg/jdn2_19.pdf?ver=2019-12-20-093655-890.

8. For more on parliament's role in strategy, see Hari Bucur-Marcu, "Financial Planning within Defense," in *Essentials of Defence Institution Building*, ed. Hari Bucur-Marcu (Vienna: Austrian National Defence Academy, 2009), 22.

9. Ibid., 55.

10. "The National Defense Institute," on the Republic of Tunisia's Ministry of National Defense website, accessed May 14, 2021, http://www.defense.tn/linstitut-de-defense-nationale/?lang=fr.

11. "What Are Dilou, Ben Ticha, Seddik, and Ghedira Doing at the National Defense Institute?" *Leaders*, accessed May 14, 2021, https://www.leaders.com.tn/article/20901-ahmed-seddik-noureddine-ben-ticha-jalel-ghedira-et-samir-dilou.

12. Interview with Hatem Ben Salem, director general of the Institut Tunisien des Etudes Stratégiques (ITES), Tunis, March 19, 2017.

13. Interview with Imed Mazzouz, director general of the National Defense Institute, Tunis, March 15, 2017.

14. Ibid..

15. Interview with Colonel Malkhaz Makaradze, commander of the Education and Military Training Command, Georgian Armed Forces, Tbilisi, March 16, 2017.

16. "About Academy; Organisational Structure," *Davit Agmashenebeli National Defense Academy*, accessed June 6, 2021, https://eta.edu.ge/en/page/4/organisational-structure.

17. International Security Sector Advisory Team (ISSAT), "Iraq SSR Country Background Note," Geneva Centre for Security Sector Governance, accessed June 6, 2021, https://issat.dcaf.ch/Learn/Resource-Library/Country-Profiles/Iraq-SSR-Country-Background-Note.

18. Interview with a defense expert in Tunis, March 19, 2017.

19. Interview with Admiral Kamel Akrout, National Advisor to the President on Security and Defense, in Tunis, March 21, 2017.

20. Parliament of Georgia, Law of Georgia on Defense Planning.

21. Ministry of Defense of Georgia, "National Security Concept of Georgia,": 2012, 2, https://mod.gov.ge/uploads/2018/pdf/NSC-ENG.pdf.

22. Ministry of Defense of Georgia, "Minister's Vision 2020," 2020, 8, https://mod.gov.ge/en/page/48/minister%E2%80%99s-directives.

23. Ministry of Defense of Georgia, "National Military Strategy of Georgia," 2005, 1.

24. Adriana Lins de Albuquerque and Jakob Hedenskog, *Georgia: A Defence Sector Reform Assessment* (Stockholm: Swedish Defence Research Agency, 2016), 18.

25. Ministry of Defense of Georgia, "National Military Strategy of Georgia," 2005, 1.

26. Ibid.

27. Irakli Garibashvili, "Foreword by the Minister," in Ministry of Defense of Georgia, "Minister's Vision 2020," 2.

28. Interview with Colonel Levan Sikharulidze, head of the Planning Division, J5—Strategic Planning Department of the General Staff, Georgian Armed Forces, Tbilisi, March 13, 2017.

29. Interview with Major-General Vakhtang Kapanadze, senior military representative of Georgia to NATO, former chief of general staff of the Georgian Armed Forces, Tbilisi, March 18, 2017.

Chapter 12. Goal 10: Military Effectiveness

1. According to Robert J. Art, these are the "four categories that themselves analytically exhaust the function of force." Robert J. Art, "The Four Functions of Force," in *The Use of Force: Military Power and International Politics*, ed. Robert J. Art and Kenneth N. Waltz, 4th ed. (Lanham, MD: University Press of America, 1993), 3.

2. Ashley J. Tellis et al., *Measuring National Power in the Postindustrial Age* (Santa Monica, CA: RAND Corporation, 2000), 133, https://www.rand.org/pubs/monograph_reports/MR1110.html.

3. Ibid., 134.

4. Ibid., 136–138.

5. Ibid., 138–143.

6. Ibid., 143.

7. Eric Schmitt and Michael R. Gordon, "The Iraqi Army Was Crumbling Long Before Its Collapse, U.S. Officials Say," *New York Times*, December 20, 2017, https://www.nytimes.com/2014/06/13/world/middleeast/american-intelligence-officials-said-iraqi-military-had-been-in-decline.html.

8. Hari Bucur-Marcu, "Financial Planning within Defense," in *Essentials of Defence Institution Building*, ed. Hari Bucur-Marcu (Vienna: Austrian National Defence Academy, 2009), 19

9. See OECD, *OECD DAC Handbook on Security Sector Reform*, 21.

10. Tellis et al., *Measuring National Power*, 143.

11. John Spencer, "What Is Army Doctrine," Modern War Institute at West Point, March 21, 2016, https://mwi.usma.edu/what-is-army-doctrine/.

12. Kees Homan, "Doctrine," Clingendael Institute, September 17, 2008, 1, https://www.clingendael.org/publication/doctrine.

13. Ibid., 14.

14. Tellis et al., *Measuring National Power*, 149–150.

15. Homan, "Doctrine," 3.

16. Tellis et al., *Measuring National Power*, 150.

17. Ibid., 152.

18. Ibid., 155.

19. Ibid.

20. Querine Hanlon, "Security Sector Reform in Tunisia: A Year after the Jasmine Revolution," Special Report no. 304, United States Institute of Peace, March 2012, 4–5.

21. Querine Hanlon and Matthew Herbert, "Border Security Challenges in the Maghreb," Peaceworks no. 109, United States Institute of Peace, 2015, 19–29.

22. Hjab Shah and Melissa Dalton, "The Evolution of Tunisia's Military and the Role of Foreign Security Sector Assistance," Civil-Military Relations in Arab States Working Paper, Carnegie Endowment for International Peace, April 2020, 1.

23. Ibid., 3.

24. Frederic Wehrey, "Tunisia's Wake-Up Call: How Security Challenges from Libya Are Shaping Defense Reforms," Civil-Military Relations in Arab States Working Paper, Carnegie Endowment for International Peace, March 2020, 1–2.

25. Interviews with U.S., German, British, and French defense attachés about their respective countries' military assistance programs and levels of cooperation with the Tunisian Ministry of National Defense, Tunis, March and September 2018.

26. Interviews with students at the National Defense Institute and instructors at the Center for Military Studies, Tunis, March 2017.

27. For example, the U.S. International Military Education and Training (IMET) programs and the Regional Defense Combatting Terrorism Fellowship Program (CTFP) both require officers be sufficiently fluent in English to attend schools in the United States. For more details, see https://www.dsca.mil/international-military-education-training-imet and https://samm.dsca.mil/glossary/combating-terrorism-fellowship-program-ctfp.

28. For example, France is reportedly funding the establishment of a Tunisian military academy in Gafsa. See Shah and Dalton, "The Evolution of Tunisia's Military," 6.

29. Tellis et al., *Measuring National Power*, 150.

30. Hanlon and Herbert, "Border Security Challenges in the Maghreb," 38–41.

31. Interview with retired Tunisian military officer, Tunis, March 20, 2017.

32. Shah and Dalton, "The Evolution of Tunisia's Military," 4.

33. Ibid., 6.

34. Interview with a member of the U.S. Iraq Office of Security Cooperation in Baghdad, April 2021.

35. Ibid.

Chapter 13. Translating the Ten Goals of Defense Sector Reform into Effective Programming

1. Note that for the purposes of this guide, we have deliberately used the term "security assistance" to encompass all forms of assistance provided by donors for the defense sector, including "security cooperation," which is the preferred term for assistance provided by the U.S. Defense Department.

2. At a training session for U.S. interagency security assistance planners and program managers facilitated by one of the authors in Washington, D.C., in 2012, a participant explained that when the U.S. government introduced a more efficient procurement system in Afghanistan, corruption increased. Belatedly, he and his colleagues realized that the inefficiency they were trying to rectify actually stemmed corruption because it was so inefficient.

3. Jeff Ernst et al. *US Foreign Aid to the Northern Triangle, 2014–2019: Promoting Success by Learning from the Past* (Washington, DC: Woodrow Wilson Center, 2020), 51.

INDEX

www.ingramcontent.com/pod-product-compliance
Lightning Source LLC
Chambersburg PA
CBHW062128020426
42335CB00013B/1144